IMPRINT

Brain Function and Dysfunction

Theoretical Neuropsychology for the Lay and Academic audience

Djedefsauron
www.djedefsauron.net
anandkumar.iitm@gmail.com

ISBN: 978-81-923559-7-9

Contents

1: Abstract

Though this thesis is meant to be layman-friendly, let's avoid a literature review on the topic of Dopamine and Norepinephrine (Norepi), partly because the *true natures* of these psychologically fundamental (as we'll soon see) chemicals were not entirely divinable to modern hyper-empirical lines of attack. For now let's just say that Dopamine is a chemical found inside animals which corresponds more to their nervous systems' *dendrites* (refer Ch. 2 if needed) and *gray matter*; and the younger cousin of Dopamine, that is, Norepi -- similarly corresponds more to *Axons* and *white matter*. Why avoid a more in-depth review here and now? As is common in such poorly grasped subjects (e.g.: the recent espousal of lobotomy in formal psychiatric circles) -- it appeared better to investigate these chemicals by an alternative theoretical approach than getting stuck in modern understandings of their natures and functions. Having chosen this line of attack, a significant finding was that animal psychological machinery involves Dopaminergic and Norepic entities (one may call them "work stations") -- which, given their special signal modulating functionality, can fairly be compared to the *Boolean logic gates* which are more familiar to engineers and computer scientists than biologists. To be precise, while the Norepic station is an elegant biological "And Gate" (as it boosts downstream signal(s) *given* the satisfaction of multiple conditions, as we'll cover in more detail in subsequent chapters) -- in contrast, the Dopamine station is a relatively simpler "booster gate" which works simply by boosting the downstream electric signal which enters its obscure machinery; however, there are no comparisons among logic gates. It is more like a quasi-amplifier. Nevertheless, for purpose of simplification, let's just call it a gate. So the Norepic *AND-like gate* and the Dopaminic *booster gate* are nature's two elegantly-evolved Boolean-style *logic gates* that play a decisive role in the organism's "quantum psychological" (by that I mean *infinitesimal signal boosting*) operations (It is not out of place to add the well-known fact that, depending on many factors, neural circuits use either more Dopaminergic, or more Norepic stations).

The second-most important finding is the cause of dementias like Alzheimer's disease. If more dopaminergic booster gates exist in a given individual, i.e., if his "neurostructural ideology" is "more Dopamine-oriented, it goes hand in glove with an elevated signal density - overly strong currents in critical regions of the brain like the Entorhinal cortex, Hippocampus, and the PFC - causing neuronal breakdown due to reasons which, though unidentified, may be any of the following or their combinations:- chemical stress (like oxidation), mechanical stress (like fatigue), thermal stress etc. In other words Prof. Slutsky et al. of Tel Aviv University correctly concluded that elevated neuronal activity is the cause of Alzheimer's disease. Such activity is "natural" to brains whose neurotransmitter systems' architecture (which one might call "neurostructural ideology") is "excessively" dependent on the (evolutionarily old) molecule of Dopamine. To put it in concrete terms -- in brains at risk of Alzheimer's disease due to lifestyles of chronic stress (which one may call "high mental metabolic rate", in other words, "elevated neuronal activity"), two details are quite remarkable:

1. Too less Norepi (doing that which Norepi does) - e.g.: "loss of noradrenergic neurons is a striking feature of dementia"
2. Along with too much dopamine gate-based information control.

Two brain models, with numbers adjusted as per my theories, are:

a. Normal brains, with around 2n[1] trillion Norepic and 20n trillion Dopaminergic work stations (whose exact physiological details and subcategories are outside the scope of this theoretical investigation)

b. Brains at risk of Alzheimer's, that have less than around n trillion Norepinephrine Gates, and around 200n-400n trillion Dopamine Gates. The last number, ~ 300n, stands out as very high. Intuition tells us that it corresponds to excessive electrical activity.

[1] Of course, this n varies depending on the brain classification.

Network collapse due to localized or chronic *current surges* -- seems linked to a family of diseases like stroke, depression, and dementia -- a family related to *hyper-activity of booster* diseases like Schizophrenia, and *Norepic gate insufficiency* diseases like cancer.

Thus Amyloid by itself seems to be merely a side-product, not the culprit that causes Alzheimer's disease. Neuroinflammation is likely merely a repair reaction. So we contrast two types of brain, the former having around 22n trillion work stations, and the latter around 301 n trillion stations, the ratio being nearly 1 is to 10!

We can simplistically neglect the variations in firing rates that are bound to exist, in order to say, that, because each – dendritic simple booster gate (Dopamine-linked Microsystem/DLMS) or Norepic long-range booster gate (NLMS) activity – adds the same amount of energy to the "total in-brain energy" – *at-risk brains* have a higher energy density, or gross metabolic rate. It's a clear observation from the above numbers. A rapid burnout of a life is caused by such "overclocking" -- which might have resulted from evolutionary pressure to remain competitive by hook or by crook, and seems to have been accompanied by an evolved dysfunction of the "pain centre" in the OFC, such that the damaging elevated neuronal activities may not be felt otherwise apart from an easily-ignored sort of depression. We can even say that the patients of depression are hard-wired to (ab)use their brains quantitatively, not qualitatively -- not use as much the key Norepic modules of the brain. A higher average/gross density of current over-taxes the current-handling brain areas, causing breakdown. A career or lifestyle of neural hyperactivity, using the brain the "wrong" way (e.g.: memorizing too much instead of thinking; note the former is more D-linked, the latter more N-linked e.g.: the involvement of Norepi in the P300) -- overtaxes the brain, causing neurons to ultimately fall apart, become unable to do their tasks. Those who're vulnerable to AD would be people having a high (mental) metabolic rate: not-physically-overactive types who eat a lot but remain thin. There seems to be a link to hair loss as well. Normal brains engage in qualitative, at-risk brains engage in quantitative signalling.

2: Key points from modern literature: a review

Fig. 1: The typical psychological neuron

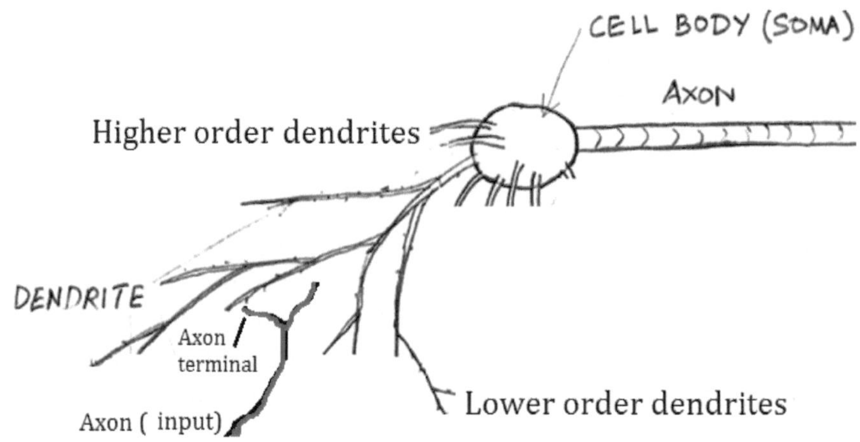

Fig. 2: The 2 Schwerpunkt *areas, here named NLMS and DLMS*:

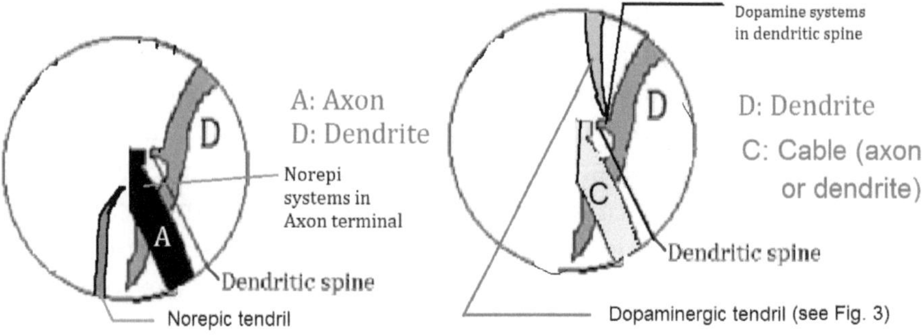

"Each dendrite may have hundreds to thousands of spines". These rapidly evolve and die out if not needed, but in adulthood spines stay for a longer time". Thus spines rise and fall as do the leaves of trees; some spinal paths are more developed, sturdier than others.

"Spines cover the dendritic tree of forebrain neurons (Cajal, 1888); it has been known for over five decades that they receive input from excitatory axons (Gray, 1959)". So, how does it work? A "Transport Axon" might bring in visual information ("red") from Eye.

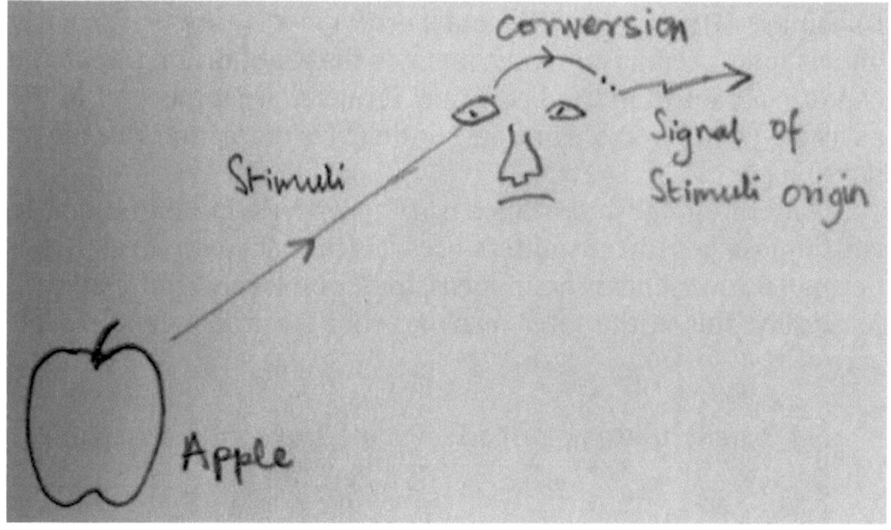

- The **dendrites** and **axons** act as signal cables, forming paths that transmit information in the form of electrical signals
- The brain's roughly 10^{11} neurons are randomly connected by roughly 10^{14} interneuron connections (synapses)
- Given how high 10^{14} is, we can say that, self-evidently, whether a particular path signals (passes current through itself) or not – is <u>in itself the information</u>; information isn't encoded in signals that're modulated.
- Thus, how a leaf detects light, a dendrite's spine detects currents
- A neuron can have maybe a thousand spines (which frequently contain the Dopaminic booster gates). Also, any incoming signal cable may feed scores of dendritic spines.

When current signaling *red* combines with other configured neurons bringing *shape information* etc. – by such timely, waxing and waning transmissions, we get the sensation "apple" – a rough example to exhibit the psychology of the Central Nervous System (CNS).

The three so-called "Classical Monoamine neurotransmitters", Dopamine (D), Norepi (N), and Serotonin (S), are particularly important. Norepi is more in control in the Right Brain; Dopamine is more in control in the Left Brain. D and N are important in the psychology of the CNS while Serotonin (S) is more important in its physiology.

Why the great importance of D and N? Define the milieu in which these neurotransmitters occur as the "Schwerpunkt" (from German *Schwerpunkt*: "main focus, focal point, center of gravity"); physically, this is the inter-neuron zone; the transmitter travels across it.

Fig. 3: Generic example of a Dopamine-Linked Micro-system

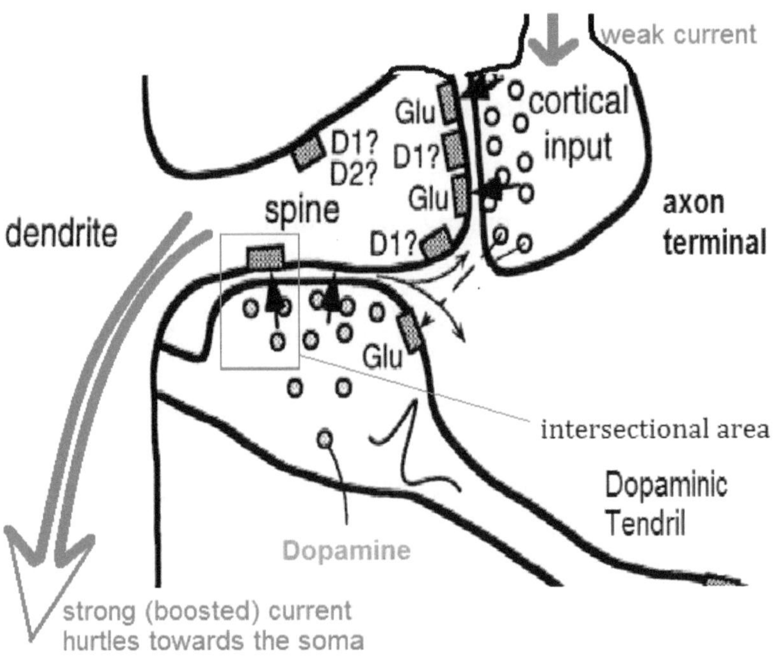

The scope of the inter-neuron zone includes the neuron sections (transmitters and receptors) between which the N or D molecules travel in "task-cycles", bringing about boosting activity within the cable that harbors the points of re-entry point of these transmitters.

Transmitters in coupled biological systems

The inter-neuron zone is a classic example of a coupled bio-system. In the context of coupled biological systems, the journey of the "transmitter" in the intersectional area is of much import and gravity.

Considering each of the two neuronal sections that together comprise the inter-neuron zone - as a separate system unto itself - and taking as analogue, for this demonstration, *the sperm cell travelling across the inter-body area between the male and female bio-systems* - as the example of this event (*impregnation) resulting in a major biological event* (that is, pregnancy) - suitably reveals - in the context of "coupled biological systems", the *transmitter's journey in inter-section area* generally is of much importance and meaning.

In this case, the meaning is signal boosting activity, which is triggered, as Nieuwenhuis will point out, by transmitter activity across the zone. This, in itself, is a valuable insight which is often overlooked.

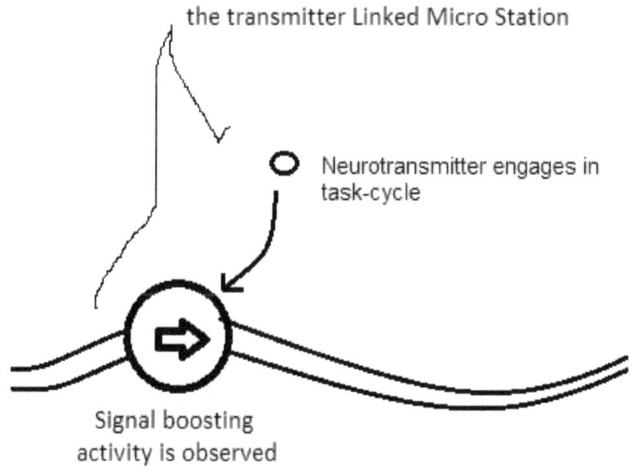

Of particular importance are the largest neurotransmitters in the inter-neuron zone – D (Dopamine), S (Serotonin), and N (Norepi).

The two major types of synapses are:

1. The dendrital area (typically a spine, a dendrital spine) type synapse, which (usually, always, or sometimes?) plays with D.
2. The Axon terminal type synapse, which usually plays with N.

We'll find that, when D passes through the zone, the structure in which it lands is undertaking 'one-to-one' signal transmutation, and for N, it is undertaking a 'many-to-one' transmutation of brain signals.

As a dendritic spine grows, what can be called the *Dopamine-Linked Microsystem* (DLMS) seems to grow in it, participating in the process described above. Similarly, NLMSs sprout along Axon terminals.

The biological or chemical Hows can't be discussed, but we can discuss the signal boosting-related Hows and the implications thereof.

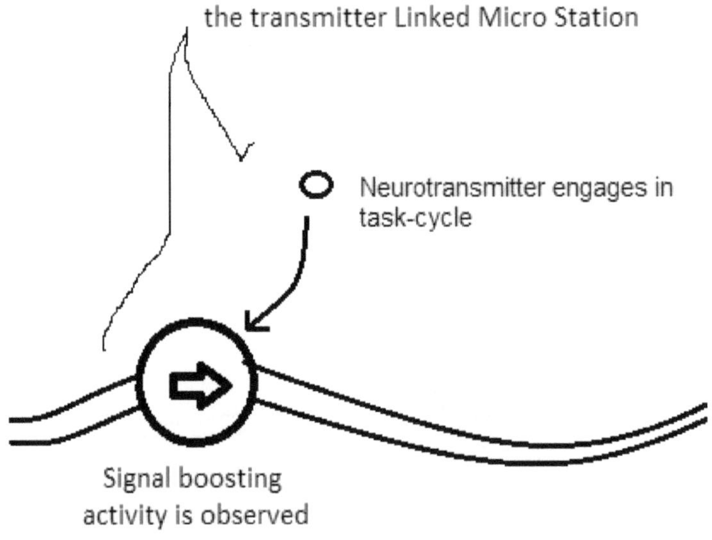

How do the DLMS and the NLMS work?

Figure 4: Dopamine and Norepi

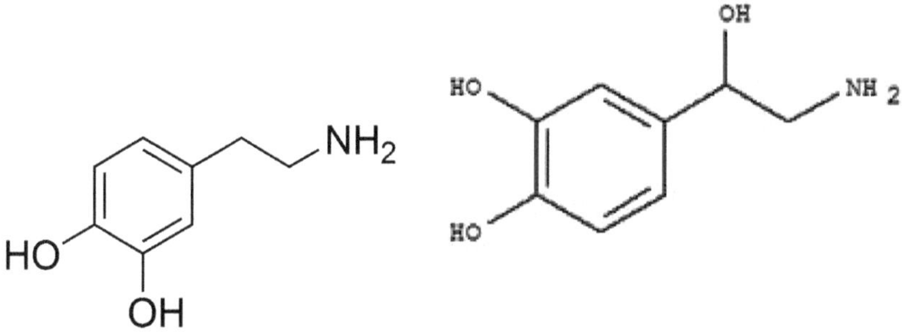

D and N are structurally and functionally similar. It follows that the DLMS and NLMS are structurally and functionally similar. They both work by **boosting (valid) signals**. As signals march through the brain, they either weaken and fade away or are boosted (energy is added to them) i.e. amplified; this boosting, i.e., amplification – is the role of the DLMS/NLMS; about this boosting, it is said:

- "Intracortical currents are triggered by the release of neurotransmitters"
- "Neuromodulators like noradrenaline, dopamine or serotonin have indirect modulating effects" - Frodl-Bauch et al., 1999, as quoted in Nieuwenhuis, 2005.
- They "enhance the synaptic responses of cortical neurons... increasing the gain of cortical neuronal activity", thus (Norepi) "serves to amplify signal conduction" (Nieuwenhuis, 2005).
    - ✓ Self-evidently, amplification of signal is one of the NLMS's major tasks. Similarly for the DLMS, as D/N are structurally & functionally similar.

So we can say that the DLMS or the NLMS has signal-modification-related roles.

That is, "quantum psychological" roles.

- The DLMS/NLMS gates boost cortical inputs, setting up the definition of behavior. What signals are seen by these gates as valid?

3: Two Types of Signalling

Signalling characteristics of the NLMS and the DLMS

We study:

Monoconditional signalling in the Dopamine (D)-linked micro-system (DLMS)

Multiconditional signalling in the Norepi (N)-linked micro-system (NLMS)

In this chapter, we shall prove that Dopamine (to be precise, the DLMS) *operates if some particular* stimuli (i.e., its signal) is present; and Norepi/NLMS operates if *more than one* signals are present.

Assumption: If D is in transit in an organism, then mostly it means that DLMS activity is on-going; similar assumption is made for Norepi.

The boosting characteristic in the DLMS and NLMS:

How the DLMS is "receptive".

Statement to be proven:

The DLMS works in a certain stimuli-reactive style (which can be called "receptive"); but the O/NLMS works in a totally different style (call it "generative"), not reactive exclusively to any given stimuli.

The *stimuli-reactive* nature of the DLMS is proven empirically; in the next exhibits, see the consistent presence of characterizable *stimuli* (which are associated with what we can call *"DLMS task-cycles"*).

- **Exhibit 1:**

Dopamine is involved with browning (Mayer, AM, 2006) in response to injury. **Injury felt by a fruit** is a *stimuli*; occurrence triggers Dopamine activity, a reaction to stimuli. Such presence of *stimuli*, in the backdrop of *DLMS activity*, will be a most consistent observation.

Evidently, both are related – the first, injury (**stimuli impinging on organism**) triggering the cycle in which Dopamine is involved.

- **Exhibit 2:**

"Dopamine is released by the algae *Ulvaria obscura* as part of an anti-herbivore defence mechanism" (Van Alstyne et al., 2006). Here, herbivore attack is a *stimuli*, which was experienced by the algae. This experience, in the form of a signal, was boosted by DLMSs and directed to a **reaction**, initiating which was the role of quasi-DLMS activity, which thus again features as a *reaction to stimuli.*

The first 2 exhibits studied simple plants, in which the DLMS exists in a rudimentary form, and maybe can even be called quasi-DLMS -- yet it retains the stimuli-reactive attribute. In complex creatures also, activity of D is observed to be featuring as reaction to *stimuli*:

- **Exhibit 3:**

Dopamine activity is observed when one is aroused; approach to arousal is, once again, a *response* to external objects translated into *stimuli.*

Dopamine activity is proven to be occurring as responses to various *stimuli* like sexual opportunity, food, money... And these are external objects felt as *stimuli* impingent on the creature. It is again clear that the DLMS activity is occurring as reactions to such stimuli.

- **Exhibit 4:**

The Spinal Cord's *diencephalospinal dopaminergic system* was discovered.

It handles "knee-jerk" auto-reactions to *stimuli*: automatically withdrawing from heat etc. Once again, D/DLMS, and stimuli to which the DLMS automatically reacts, are seen as occurring side by side.

- **Exhibit 5:**

The Sponge

One of the best sources of empirical data about the behaviour of the DLMS is the sponge, because it is a pure D-S nervous system (Chapter 5). Now that we have found Dopamine in sponges (Liu et al., 2004)...

1. "Most sponge species have the ability to perform coordinated contractions squeezing the water channels and thus expelling excess sediment and other substances that may cause blockages [the presence of blockages were ***stimuli*** that caused a ***reaction***].
2. Sponges contract to reduce the area vulnerable to attack [a *response* reacting to "attacker", presence of which is again a *stimuli*]. The mechanism is unknown, but may involve chemicals similar to neurotransmitters." The mechanism involves DLMS activities in the sponge's "D-S" nervous system (Chapter 5).
3. In fact, so dependent on stimuli-reactivity is the sponge that it is said: "Sponge lacks nervous, digestive or circulatory systems; instead the water flow supports all these functions".

Using this, we substantiate the statement i.e. DLMS activity is "receptive", as DLMS activity occurs, in the main, as a response to stimuli; and throughout its life-cycle, the signal is based mostly in dendritic dendrital areas – that is why we label all such activity as "receptive". "Generative" signals use more the axonic power of Norepi.

NLMS activity is "Generative", not simply stimuli-reactive

Statement to be proven:

> Though the DLMS is involved in reactive behaviours, the O/N[2] system is involved with more than just knee jerk-type reactions to stimuli. Some exhibits are studied to bear out this point.

Exhibit 1:

"In locust jump, Octopamine (O) modulates muscle activity, making the leg muscles contract more effectively". Jumping of the locust doesn't appear to be a direct reaction to any recognizable stimuli (as it is with Dopamine). It may be a reaction to more than one stimuli...

Exhibit 2:

Octopamine "is widely used by invertebrates" in what are called "energy-demanding behaviours", like flying, egg-laying, and jumping...

These behaviours, which involve O/NLMS activity, are not dependent on any one observable stimuli, as was in the case of D. They may even be characterized as novel, not automatic/stimuli reactive.

Exhibit 3:

"In the firefly, Octopamine is associated with light production in the lantern".

A pattern is seen:

O/N is involved with "contrived" actions, not direct responses to stimuli.

---

[2] In Chapter 5, we see how, while N and O are physiologically different, they are psychologically similar. Hence we can cite examples now of both OLMS and NLMS, as the psychological roles of both are the same.

It may seem peculiar, as Dopamine and Octopamine/Norepi activity must be similar to each other, given the similar molecular structures. In fact, on further inspection, it appears that -- in contrast to how 1 incoming signal causes reaction in the DLMS -- in case of the O/NLMS activity, **more than one signals should be involved**.

To explain that point, we may deconstruct how a fire-fly puts on its lantern.

This activity depends on, say:

a. *Whether or not it is night* (a TRUE or FALSE value, conveyed as "a signal" or "no signal" incoming in the signal cable)
b. *Whether or not a female is nearby* (a TRUE or FALSE, conveyed as a signal or no signal incoming in the signal cable)
c. And/or some other data, such as *whether or not there is enough fuel* (a TRUE or FALSE value, conveyed as signal or no signal).

All three must be TRUE; only then will the O task-cycle occur and light come on. So the O/N-linked micro-system seems like an 'AND gate'.

A simplified, two-input version of such an AND Gate is shown below:

Figure 5.1: Seeing the NLMS as an AND Gate

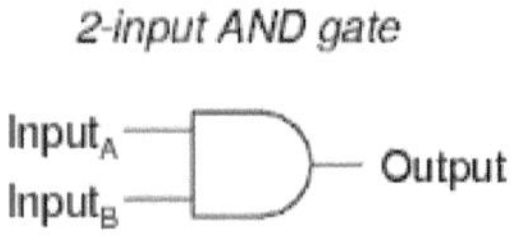

2-input AND gate

Input$_A$, Input$_B$ → Output

A: It is night
B: Female is nearby

1 = True
0 = False

A	B	Output	
0	0	0	**lantern off**
0	1	0	**lantern off**
1	0	0	**lantern off**
1	1	1	**lantern on**

Exhibit 4:

In accordance with the above way of looking at things, we see that N, the vertebrate's psychological equal of the invertebrate's Octopamine, is similarly found involved with reaction to multiple stimuli!

So we see:

Chapman and Bragdon, 1964 discuss the "P300", a specific kind of neural wave pattern later found to involve Norepi activity. They use the term "*meaningful*"; one stimuli (signal set A) can be meaningful only in the *context* of something else (signal set B). It appears the NLMS deals with multiple signals, as in the case of the firefly!

Proving the presence of multiple incoming signals is a simple task:

"In later studies published in 1967, Sutton and colleagues had subjects guess whether they would hear one click or two clicks. They again observed a positivity [P300] *around 300 ms after the second click occurred -- or would have occurred, in the case of the single click".*

Figure 5.2: Seeing the NLMS as an AND Gate

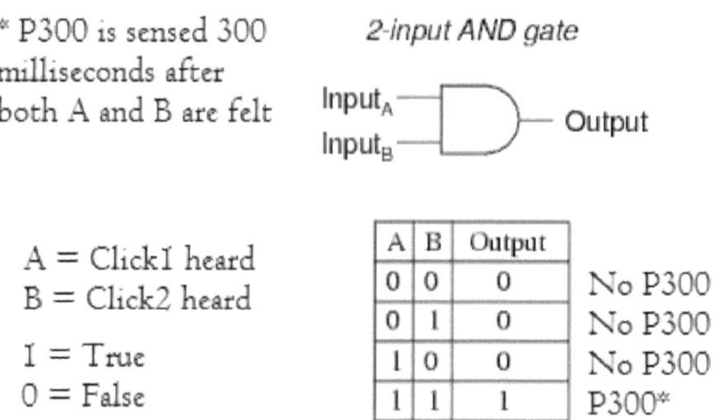

A	B	Output	
0	0	0	No P300
0	1	0	No P300
1	0	0	No P300
1	1	1	P300*

c. "The P300 is thought to reflect processes involved in stimulus evaluation/categorization". "Evaluation/categorization" implies stimuli being weighed *in the context of certain other data* – which again proves the P300/NLMS activity's multi-input (signal) nature.

Thus the Norepi-task-cycle calculates the output from various signals.

Of Norepi's many-in, 1-out type of signalling, Nieuwenhuis et al. say:

> “NE [N] release is elicited"... "Collapsing the vast information processing circuit to the outcome of a single decision layer.”

Thus the difference is between D's simple stimuli-reactive i.e. *monoconditional* signalling, and N’s *multiconditional* signalling. This is far more interesting in the light of this chief detail: Norepi task-cycle sites are, usually, axons; Greengard et al., 1984: "Widespread effects of noradrenaline [Norepi] on axon terminals". DLMS task-cycle sites are dendrites, except in special cases; Yao et al., 2008: “Dopamine receptors are present throughout the soma and dendrites, but ultra-structural and biochemical evidence indicates they are concentrated in dendritic spines".

4: D/N Logic Gates in Neural Circuits

This chapter summarizes the lessons learnt from chapters 2 and 3.

One can pictorially summarize how the NLMS works, as shown below:

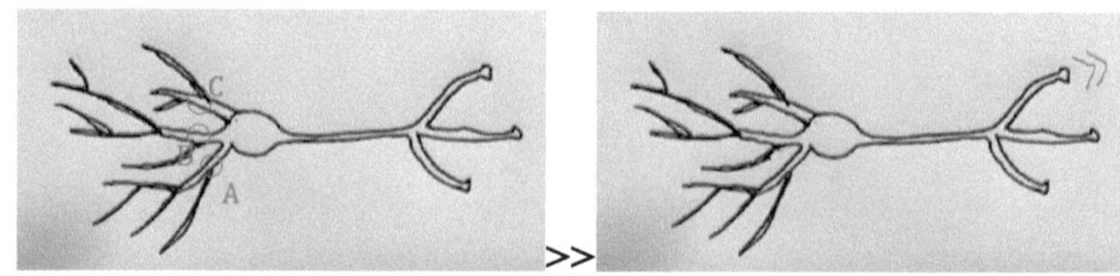

(Relevant set of signals at *n* **higher order** dendritic sites leads to boosting at the NLMS site at the axon terminal head, and current goes forth).

One can pictorially summarize how the DLMS works, as shown below:

(Relevant set of signals at one **lower order** dendritic site leads to boosting at the DLMS site found in the near vicinity, and current goes forth).

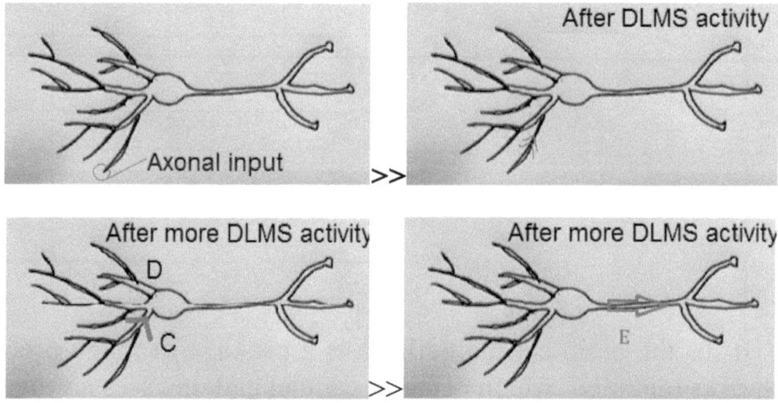

(The axon transmits signal because the sheer strength of the signals coming from, say, C and/or D – *rams* the signal through E in "action potential method" – the axon doesn't fire "of its own will", using the Sheffieldian mechanism; the axon merely serves as a signal cable).

In the NLMS-guided mode of transmission the axon fires because the NLMS feels (somehow) the weak dendritic signals at A, B and C; the axon fires "of its own will". Axon is a detector, not just a signal cable.

Electronic circuits provide a handy analogy to understand how Dopamine-gates (DLMSs) and Norepi-gates (NLMSs) are variously permuted to form neural circuits which variously process information.

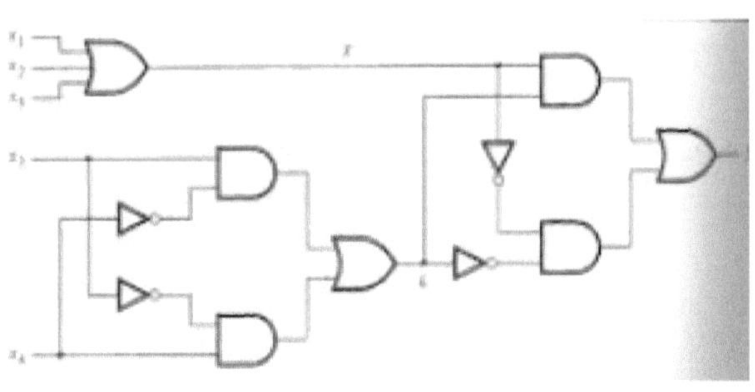

monoconditional signalling

A DLMS gate

A	Output
0	0
1	1

multiconditional signalling

A B C NLMS gate

A	B	C	Output
0	0	0	0
0	0	1	0
1	0	0	0
0	1	0	0
1	1	0	0
0	1	1	0
1	0	1	0
1	1	1	1

Thus the brain can be defined as a package of many processing networks (circuits), which occur in parallel and/or series arrangements – each circuit is made up of Dopaminic and Norepic signalling work-stations.

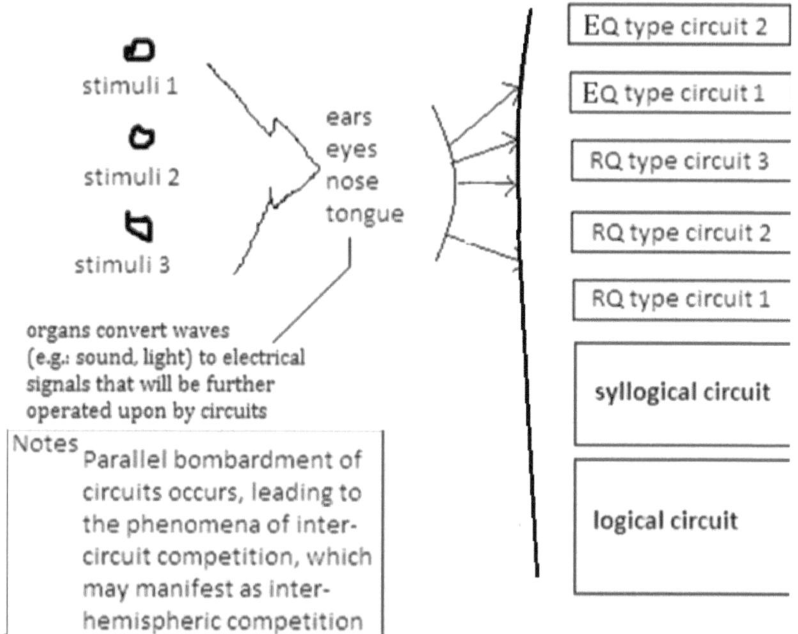

Traditionally it was understood that dendritic spines collect electric signals, after which these signals cumulatively add up, and are then rammed through axons, as shown in the below picture. This mechanism forms the basis for the *perceptron* neural networks.

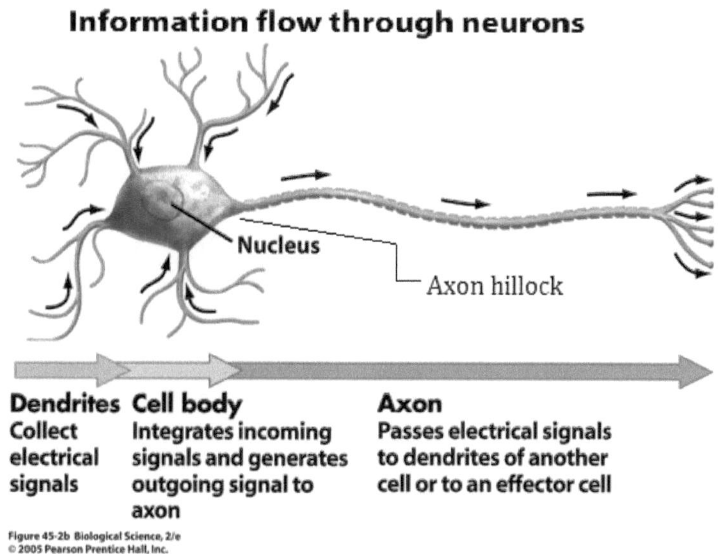

But, in complete harmony with the research discussed until now – Sheffield has discovered a new type of Signalling: "Sheffield et al. assert that some action potentials began at the distal end of the axon (the end not connected with the cell body) instead of at the axon hillock", and consequently, that "axons can communicate with each other without the signal first going through dendrites or cell bodies." The NLMS explains this axon-centric transmission, evidently.

But even in this (Sheffieldian) case, there should be signals in the *dendrital* side, albeit they were perhaps small and consequently, not easily measurable – the reason for this statement is that an "**Axonic**" neuron's axonal elaboration is accompanied by more "higher order" (Fig. 8) dendrital elaboration, as proven by how Chris Chatham says: "The left hemisphere [*where dendrital elaboration is more* (*and the opposite trend holds for the right hemisphere*)] has larger dendrital branching than the right hemisphere, but only at large distances from the dendrite's main shaft – the opposite trend holds at distances closer to the main dendrite" [so the "axonic" neurons of the right hemisphere have more "higher-order" dendrital branching; this appears to be a Sheffieldian-transmission-oriented neuronal design, which we term as "axonic"].

Figure 8: How axonic and dendritic neurons differ

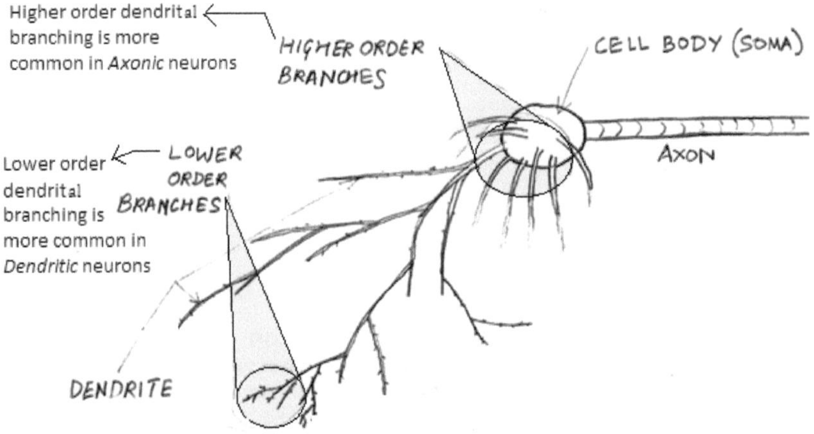

Thus the "quantum episode" is about the DLMS or NLMS's modification of neurodata; how this modification follows - in case of either D or N - consistent logical laws. It's about how the Micro-System reacted, or, converted - weak signals (*signals of stimuli or processed stimuli origin*) to another logically following neurodata; the question is, as you have already pointed out, not about an advanced neural control system "homunculus", *which gives input to LC-NE modulatory system* - it's but a misleading idea seemingly inspired by the widely criticized idea about the "Dopamine reward system". The false view imagines the neurotransmitter source (for example LC/SN) is a *system that releases N* (or D) *according to psychologically relevant inputs* - but it is far more likely that the LC/SN is merely a refilling system of absolutely no psychological importance. When the Locus Coerelus generates Norepi, and feeds it onwards to the site -- it is not on the basis of any other input apart from the input provided by the NLMS that says, "My Norepi molecules were spent in local operations, refill me"!

Thus, surely, if one asks, "What makes a brain different from the other brain"? The answer seems to be **the layout of D and N "check-posts"**, which are arranged in different permutations and combinations for different behaviours. If there is any fundamental mental coding which one can analogize to DNA, it is this D/N layout (much of which is changeable). There are hundreds if not thousands of different behaviours expressible as calculative D/N circuits, the majority of which may be similar from human to human.

This style of the DLMS or NLMS is used by signal-processing *neural circuits* - both *more D-reliant, less N-reliant* circuits and *more N-reliant, less D-reliant* circuits. Such are the circuits which define personality, which is all about how the myriads of DLMS or NLMS are arrayed, so a man behaves in specific ways depending on that.

5: The Mysterious History of the CNS

This chapter gives solid grounds proving the claims made in chapter 4. As the Soviet biological scientist Dobzhansky said: "Nothing in biology makes sense except in the light of evolution"; and this seems to hold truest in case of the **classical monoamine neurotransmitters**. Any "minor" change in the Schwerpunkt caused a disruptive evolution simply because of its primary importance.

Life forms are ultra-complex chemical reactions (i.e., **super-reactions**); the mentioned neurotransmitters are "stable ranged chemicals" defining **super-reaction**'s deepest aspects (reactions taking place across the longest range; for example the Dopamine molecule may begin at the SN and end up in the PFC, among other areas -- which other chemical in the CNS might travel such a long distance?)

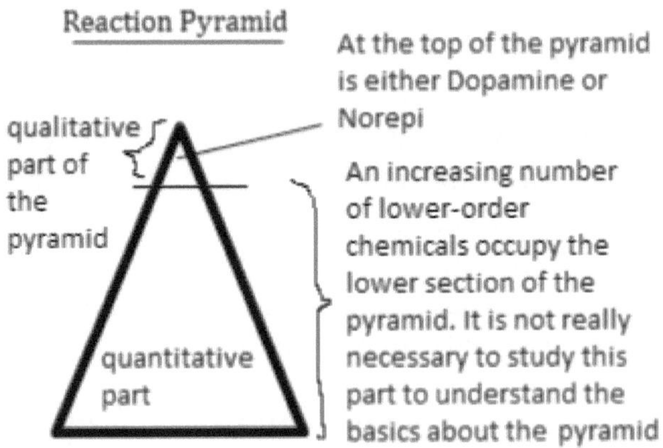

The far-going D and N – comparable to the sperm in *analogy: impregnation* -- are central components of the Schwerpunkt. Studying their evolution may help us understand the evolution of life. Must get a Eureka on seeing that, for example, Octopamine is present in *all* invertebrates; Norepi is present in *all* vertebrates. It must occur to you that vertebrates are vertebrates *because* Norepi, and not Octopamine, is present in them, rather than this detail being merely some corollary of some deeper element of life design.

The very fact that Octopamine is present in invertebrates and N in vertebrates -- should hint that modification, or evolution, or mutation of these neurotransmitters -- may have been the **fundamental change** underlying the evolution of life to greater complexities.

That "small" change was but the butterfly's flurry which causes the storm, the manifest differences between vertebrates and invertebrates – and the same thumb-rule differentiates, say, plants and animals. In other words, the rise of new, differently behaving classical monoamine neurotransmitters was at the root of things, as one class of life-forms evolved into another. Thus, to say that D evolved into O and O evolved to N, is the premise of this thesis.

Figure 6: Evolutionary timeline

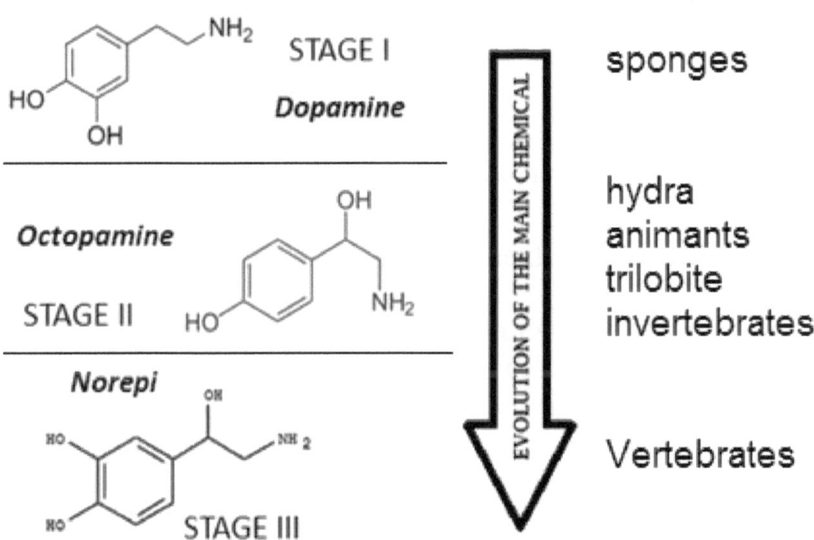

The Plant

See the humble *plant,* for example, where the basic psychological neurotransmitters D, O, or N, are not present in that "chemical soup" or substrate – in which they usually occur in higher life-forms.

To be sure, Serotonin is found in plants, serving up a basic chemical substrate that enables plant processes. But this chemical soup does not serve to D, O, or N. So we say that the plant is simply a *S base of reactions*, rather than having evolved to what we can call the *D-S base of reactions*, which is the next step of complexity.

(Base of reactions can be defined as the **central class of chemical reactions** in the multifaceted chemical reaction called "life-form").

That is not to say that D is *not* found in plants -- both D and S may be found in plants (for example, D features in the activity of fruit browning) -- but they don't occur in the relevant "D-S system" type of nervous system -- a highly specific set of D-S interactions seen in life-forms of the post-plant rung of complexity.

Of the plant, the *"S base of reactions"*, let's see what became of it.

The planimal

We all originated from a plant-like parent wherein the **S base of reactions** reigned. Of course, this transformation happened in stages.

The first stage involved the rise of a clumsy creature we may call "planimal".

The causal event behind the rise of the planimal was the rise of the critical **D-S base of reactions**; it was a disruptive evolution, occurring in the Schwerpunkt, and thus totally transforming the life-form.

It is well-established that S modulates D systems (SL Dewey, 1995), which is insight into the D-S base of reactions. D is a simple molecule relative to the similar Norepi, hinting that the *D-S base of reactions* (which features D's role as a neurotransmitter) evolved first, as an old CNS (before the rise of the O-D-S base of reactions).

The **planimal** was the first post-plant, followed by the versatile, Octopamine-blessed **animant**, and only then came the animals.

Good news is that these ancient creatures still exist among us, in the very form in which they were before nearly all life on earth began! The planimal, a life-form defined by a D-S base of reactions – evolved when the available plantish molecule of Dopamine (D) entered a specific relationship/reaction with S – thus entitling, to D, the privileged status of 'neurotransmitter', starting the career of the proto-CNS in the planimal, whereby cells collaborated in a lifestyle more complex than that of plant; it was the first quasi-animal. An example is the *sponge* of the ocean, in which have been found both Dopamine (Liu et al., 2004) and Serotonin (Weyrer et al., 1999).

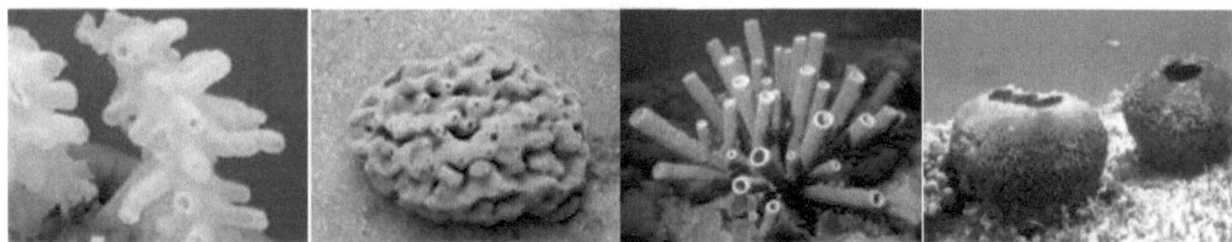

The sponge's purely stimuli-reactive behaviour (chapter 3), which distinctively evidences the activity of Dopamine, supports the view that the sponge is home to (no more than) the D-S base of reactions.

The sponge has long been suggested to be the "missing link" between plant and animal; now it proves true. Another proof that sponges did in fact develop from plants is how some sponges are photosynthetic.

The animant

Octopamine, the invertebrate molecule, evolved before N, the vertebrate molecule. In the next disruptive evolution, either D evolved into O, so beginning the unstoppable march of the Fallen Angels.

Or O, a molecule found in certain plants (or indeed planimals) was "captured" by the "chemical soup" – the O-S-D (or D-O-S) base of reactions emerged. Some of these possibly "accidental mutations" were "biologically successful", i.e., some O-involving reactions were accepted by the old D-S base of reactions; entropy allowed it.

O-D-S implies the primacy of O in the total reactions pyramid; and D-O-S implies the primacy of D in the total reactions pyramid. Why this bifurcation? We have reasons to think so, to which we'll return.

We talk about "primacy" and "rivalry" in the context of O/N and D, because there are, in the quantitative part of the pyramid, chemicals which can react with *both* D and N, which become, in a sense, like rival suitors. Thus the rise of O-S sprouted two types of creatures. This D-O-S or O-D-S creature may be called *animant*; an example is the hydra (which, like plant, enjoys, in theory, infinite lifespan).

The activity of the O-class chemical's classic psychological property of calculation is readily seen in how "Hydra are usually sedentary, but occasionally move quite readily, especially when hunting".

The rise of O introduced the calculative (multi-conditional) abilities covered in chapter 3. Another key aspect of the animant, a corollary of such calculative skills, was the presence of a certain "learning ability" expediting organ development. And thus, while the "Sponge [the planimal] lacks nervous, digestive or circulatory systems" -- i.e., they lack true organs tailored for the specific job – the meta-O-S creatures, the 'animants', began organ development; the brain, the limbs etc., may have been some of the first organs developed. Some say "Hydra does not have a recognizable brain", but the presence of radial symmetry is proof that it has some sort of "brain".

Octopamine had a talent of using the O-involving process to relay orders *discriminatingly* to other cells – those then gladly adapted to the leadership thus provided, taking on a myriad of roles.

That brought about specialized organs. Because, for organs to work, discriminatory (multiconditional) signalling is required; organs can't function at the "always on", directly stimuli-reactive mode.

We can refer to this organ-growing property of O/N by the term "localized evolutionary intelligence"[3]. It is absent in sponges, which "lack nervous, digestive or circulatory systems". Proves the idea of the sponge as a humble planimal, with only a D-S base of reactions!

The "organ master" role of O/N is apparent: N "triggers the release of glucose from energy stores, increases blood flow, heart rate, oxygen supply etc. in the fight-or-flight situation". That detail should not cloud our understanding of its psychological centrality, however.

Similarly, Axons are the taskmasters of organs: "At synapses, axons make contact with other cells, usually other neurons but sometimes muscle or gland cells", ordering with discriminating ability.

The Axon/O/N's organ master role is confirmed in how injecting O into "a lobster and crayfish resulted in limb/abdomen extension" (Livingstone et al., 1980), i.e., caused O booster gates' activity.

The Bilaterally Symmetric animal

"Sponges are like other animals: multicellular, heterotrophic, and lack cell walls [a key aspect of plant's S base of reactions]; and produce sperm cells; but unlike other animals, they have no body symmetry."

That planimals have no body symmetry -- is interesting in the context of symmetry. Clearly, the next disruptive evolution was the rise of bilateral symmetry, which is found in invertebrates like flies also.

---

[3] Octopamine is found in plants that show 'localized evolutionary intelligence' e.g. *bitter* orange (Tang, 2006), but lacks the chemical soup.

Just how certain planimals retained plant characteristics (like photosynthetic sponges) – some animants could still *fuse* like the sponge[4], it is evident, given this theory about the next disruptive evolution:

Figure 7: Rise of the hemispheres

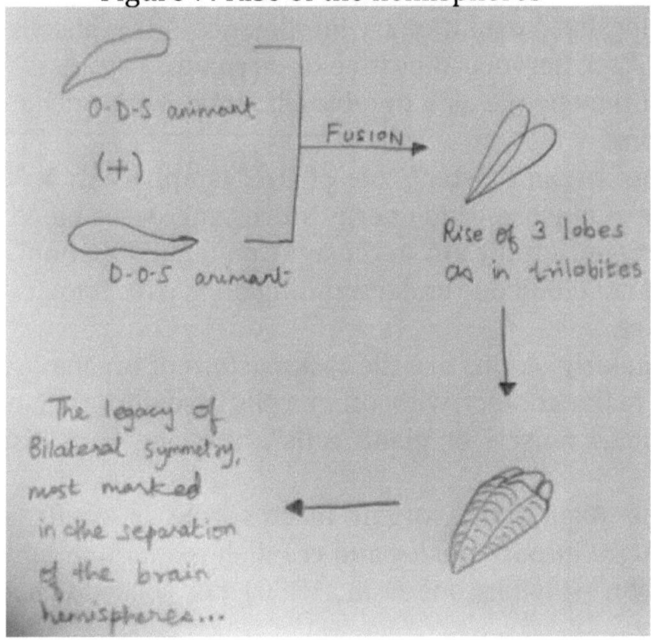

Thus a pre-trilobite creature emerged as the fused D-O-S + O-D-S creature, the two-brain invertebrate animal that had bilateral symmetry.

As far as life-forms go, it was a most successful design; even we are bilaterally symmetric! "Animals" may be defined as those who are bilaterally symmetric (so they're associated with a touch of unpredictability, hypocrisy etc.); a few animants like Jellyfish and hydra are radially, not bilaterally symmetric, showing that either the O-D-S or the D-O-S super-reaction is in them, but not both.

---

[4] Two sponges may even "fuse" with each other, and "average out" their traits; "When two sponges are fused, contraction waves slowly become coordinated." Some such thing may have caused the rise of male and female in animants, as sponges, though having sperm cells, are hermaphrodites

Evidently the Left Brain is the D-O-S system (later became D-N-S), where D has primacy of some sort, as reflected in what we know of it; and the Right Brain was the O-D-S system (and later, N-D-S).

Appearing shortly before the trilobite, *Vernanimalcula* is the earliest Bilateria (fused, symmetrical); from 550 million years ago (Chen et al.). It was "thin like hair". Unsurprisingly, soon after, the earth paid witness to the bi-brained invertebrate 'trilobite', which was an unprecedented evolutionary success, as your digs amply show: The trilobite became as typical as to become the geologist's means for the measurement of Pangea landmass, plate tectonics etc.

Rise of the decentralized nervous system

There was, further, another disruptive evolution when O was replaced by the similar N (Norepi). Merely on the basis of how *vertebrates have Norepi whereas invertebrates have Octopamine,* we can say that N brought into the picture a new physiological ability for the nervous system to be decentralized, as borne out by the presence of a spinal cord (which is the main detail, beyond the vertebra). Otherwise, psychologically O and N are similar. The difference is physiological, as best seen in the centralization of the octopus.

6: Theory of D/N Subsystems in Body

Next presented is a theory regarding a *Dopaminergic System vs. Noradrenergic System* **functional dichotomy**, which is seen across **pairs** of similar organ systems adjacent to each other in certain body areas?

Pairs of superficially similar-looking, but functionally slightly different, Norepi-based and Dopamine-based biological systems, in the ratio of maybe around 20:1...? (20 D-structures for every N-structure?)... Both N-based system and D-based system exist side by side.

The D-based system is a less important but relatively plentiful *evolutionary reflection* of the majorly important N-based system (just how the left brain is basically a Dopaminic reflection of the right).

Examples:

1. In the neuron (Norepi-based axon, Dopamine-based dendrites)
2. In tongue (Norepi-based fungiform papillae vs. Dopamine-based[5] Filiform papillae)
3. In the photoreceptor (Possibly N-based[6] **cones** vs. D-based rods)

If this theory is correct, certainly more cases can be found by the scientific community. Anyway next we'll examine all 3 cases in detail.

The scariest eye diseases are linked to cone malfunction or deficiency.

The functional differences are such that, in every case, while the D-based system can be called quantitative, the corresponding N-based system can be called qualitative (words you should use, here).

---

[5] That Filiform papillae are associated with Dopamine is another prediction thrown up; researchers can verify whether this is true or not.

[6] "Cones are associated with Norepi" – it is a prediction thrown up by this hypothesis... Researchers must verify whether this is true or not.

The Axonic and dendritic systems have differing memory and processing roles, as we'll see in later chapters. The NLMS is at the *heart of the axon terminal*; and a similar statement about the D-based biological system (DLMS) at the *heart of the dendrital spine etc.*

Now, by comparing axons and dendrites, we will substantiate our statement that the N-based system is qualitative, possessing enhanced functionality and cooler functions, although as always few in number. And the D-based system tends to be "quantitative" (*limited functionality and many in number*). Similar comparisons will be carried out for the cones and rods, and for the two types of papillae.

To be sure, axons and dendrites are superficially similar, e.g.: both conduct electricity. But the differences between them are many.

	Axons	Dendrites
Axons score lesser on quantitativeness; see, for example, their quantity	Single axon per neuron	Many dendrites per neuron
Quantity of branches is the parameter studied:	Fewer branches	Many branches
Axons score higher on the scale of qualitativeness; first structural quality to be compared, is length:	Very long (may be several metres)	Very short (generally under 1.5 mm)
An aspect of structural quality is the uniformity of diameter	Uniform diameter	Tapering diameter
Yet another aspect of structural quality is the perpendicularity of the angle for new branches:	The branches of the axon are at right angles to the axon	Dendrite branches are not usually at right angles
Another aspect of structural quality:	Terminal branches enlarged to form **synaptic knobs** at the tips	No knobs at the tips of branches
Another aspect of structural quality:	Have neurotransmitter vesicles in their knobs.	Do not have such vesicles.

Thus we can generally say that the NLMS-containing Axons are qualitative; the DLMS-containing dendrites are quantitative. This qualitative vs. quantitative "functional dichotomy" seemingly has its origins in the fact that Norepi evolved more recently than Dopamine, and is just different in the style of chemical reaction; the present theory's correctness is again upheld by the following observation:

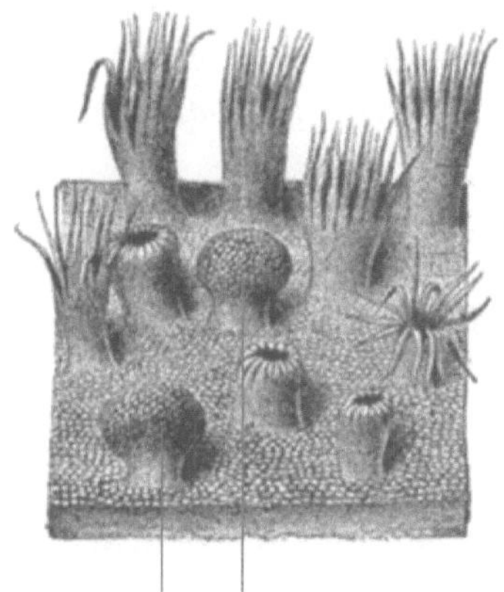

2 (tongue) fungiform papillae are shown; the rest are filiform papillae.

There are 2 main types of "raised protrusions of the tongue surface called papillae":

1. "Fungiform papillae - Slightly mushroom-shape, longitudinally; contain taste buds. These are present mostly at the dorsal surface of the tongue, as well as at the sides. Innervated by facial nerve.
2. Filiform papillae - thin, long papillae "V"-shaped cones that don't contain taste buds but are the most numerous. These are mechanical and not involved in gustation, and characterized by increased keratinization".

Therefore, we can say:

Fungiform papillae are “qualitative”, and Filiform papillae are “quantitative”...

And now we see:

"Fungiform papillae were found to contain Norepinephrine" (Zancanaro et al., 1997)... Observations like this add credit to this theory...

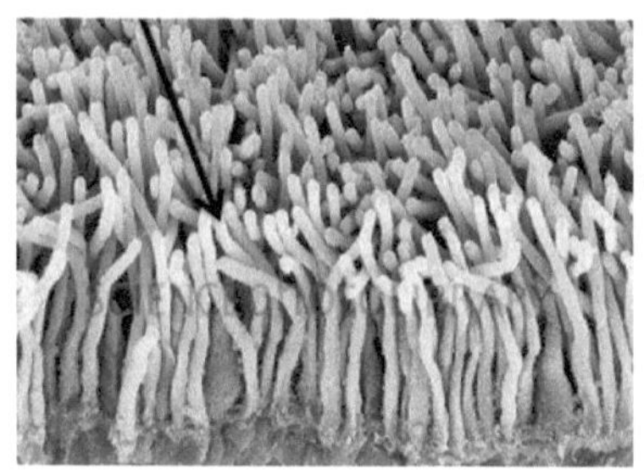

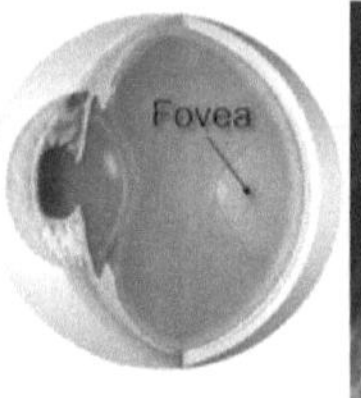

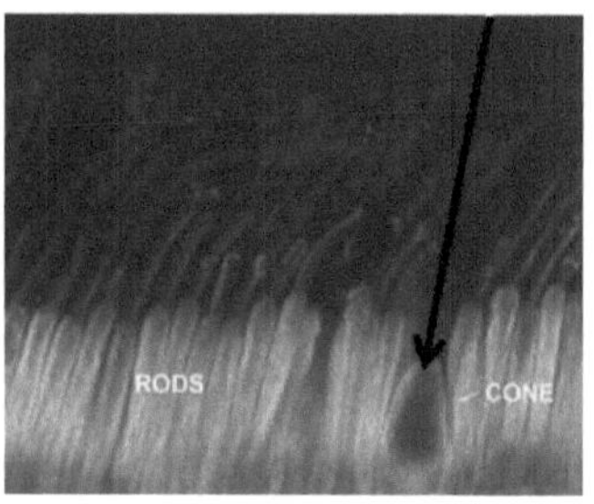

A similar functional dichotomy is seen in the retina of the eye, where cone and rod photoreceptors similarly differ. Helga Kolb says: "Photoreceptors in most vertebrates including human, have a content of D2 dopamine receptors somewhere upon their surface (Witkovsky and Dearry, 1991)" -- it is nevertheless likely that Norepinephrine, rather than Dopamine, is associated with cones...

Dear Sir,

You are better positioned to verify that than me.

Thus the main theory produced in this chapter predicts that Norepic activity underlies cone activity, and Dopaminic activity underlies rod activity. The theory uses the following fact to make that statement: Rods are quantitative, unlike cones, which are qualitative:

- "The retina has ~ 6 million cones and 120 million rods[7]" (Schacter et al., 2011);
- Rods are associated with quantitative caches of data e.g.: "Rods are extremely sensitive, and can be triggered by as few as 6 photons" (Hecht et al., 1942).
- "There are major functional differences between rods and cones."

These are summarized:

---

[7] This develops our theory that the N:D Ratio, universally, is about 1:20...

	Cones	Rods
Quantitative trait: more pigment is present in rods:	"Have less pigment than rods (so require more light to perceive images)"	"Have more pigment than cones, so can detect lower levels of light"
Qualitative aspect of cones is the pigments' variety	"Three types of photosensitive pigment in humans"	"One type of photosensitive pigment"
Rods seem to be quantitatively sensitive, unlike cones	"Not very light sensitive; sensitive to only direct light. Cones are used for vision under high light conditions"	"Very light sensitive; sensitive to scattered light. Used for vision under low light conditions"
Loss of cones causes absolute blindness; so we can say cones are involved in "qualitative" type of vision:	Loss of cones causes *legal blindness*	Loss of Rods causes the much milder *night blindness*
Qualitative attribute of cone is again obvious:	High visual acuity; better spatial resolution	Low visual acuity
A qualitative aspect is how cones enjoy preferential spatial treatment, how axons branch at 90 degrees unlike dendrites	Concentrated in fovea	Not present in fovea
Qualitative functional aspect of cones:	Fast response to light, can perceive more rapid changes in stimuli	Slow response to light **Stimuli added over time**
Qualitative aspect of cones:	Disks are attached to cones' outer membrane	No disks
Qualitative aspect of cones:	Confer color vision	Confer achromatic vision

7: The Neural Spectrum of Humanity

If deficiency of cones is possible –

- "Color blindness is caused by a fault in the development of retinal cones"
- "Destruction of the cone cells from disease would result in blindness."

– And deficiency of fungiform papillae is possible –

- Fukutake et al., 1996: "Absence of fungiform papillae on the tongue"[8].

– Then certainly, deficiency of Axons etc. is also possible. Indeed this condition where axons/Norepi/white matter is lacking is "white matter disease":

Dr. Nicole Anderson: "Our findings add to a growing body of evidence that white matter disease is a discreet saboteur in the brain, impacting a large number of cognitive functions. It is responsible for about a fifth of all strokes worldwide (Sudlow, 1997), more than doubles the future risk of stroke (Debette 2010, Vermeer, 2007), and is a contributing factor in up to 45% of dementias (Gorelick, 2011)." The prevalent view is an "*evil, villainous disease*" idea, which goes: "white matter disease is a mind-robbing condition that targets small blood vessels deep within the brain's white matter. The disease hardens the tiny arteries, gradually restricting nutrients to white matter" – however, the victim's genes/lifestyle/environment, which results in less NLMS/axonic/white matter type processes, and more of the Dopaminic – is the real culprit in WMD. White matter may wither away (the *use or lose* law) even as grey matter declares its fiefdoms in the brain.

[8] And other relevant symptoms: "Sural nerve biopsy demonstrated a marked loss of myelinated fibres [*axons/white matter*] and a reduction in the number of unmyelinated axons"... "Neuro-imaging studies revealed atrophy of the spinal cord, cerebellum, brainstem and corpus callosum and enlargement of the lateral, third, fourth ventricles."

Definitely deficiency of white matter/axons/Norepi is possible in brains; that whole aspect may weaken; it is called “Norepinephrine deficiency”.

dopamine

norepinephrine

HO

HO

NH_2

dopamine β-hydroxylase

O_2

ascorbic acid

H_2O,

dehydroascorbic acid

OH

HO

HO

NH_2

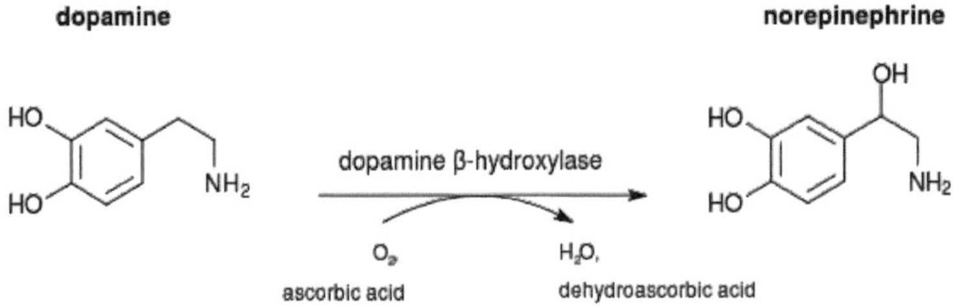

What is Norepinephrine deficiency? “Dopamine-beta-hydroxylase deficiency is a condition involving inadequate Dopamine-beta-hydroxylase. It is caused largely by [or rather, it is associated largely with] increased amounts of serum dopamine and release of dopamine in place of Norepinephrine; sometimes it is known as Norepinephrine deficiency”.

“Researchers of depression, schizophrenia, and migraines are very interested in studying this disorder”. It is associated with severe symptoms such as Ptosis of the eyelid, Prolactin deficiency (female’s inability to produce milk), and difficulty standing still for longer than one minute.

Ptosis of the left eyelid (unilateral ptosis). An 1852 headshot daguerreotype by William Bell.

A symptom is hypoglycemia – abnormally diminished content of glucose in the blood (which is because of how, as shown in chapter 10, their hyper-D Systems are quickly burning up carbohydrates, i.e., a high metabolic rate). “According to a NIH study, 100% of those [suffering from Norepinephrine deficiency] studied suffered severe orthostatic hypotension, 80% suffered anemia, 43% epileptic symptoms, 100% nasal stuffiness, 33% suffered hypoglycemia, 60% frequent urination and night-time frequent urination, 50% muscle hypotonia, 75% postprandial hypotension; 100% suffered from sleep problems”.

Quoting part of the abstract of a very important 1981 paper by Cross et al.: "Dopamine-beta-hydroxylase activity was measured. Enzyme activity was decreased in the frontal and temporal cortices and hippocampus in patients with Alzheimer's disease. This decrease in activity in AD may reflect abnormality of cortical noradrenergic fibres."

Thus in some brains there is deficiency of Norepi (or Dopamine-beta-hydroxylase), and, in parallel, excessive presence of Dopamine: "persons with Dopamine beta hydroxylase deficiency [Norepi deficiency] have **triple fold amounts of dopamine in their system**". Pagon et al., 2003, similarly note: "Biochemical features unique to DBH deficiency include minimal or absent plasma norepinephrine and epinephrine and a **five to tenfold elevation of plasma dopamine levels**".

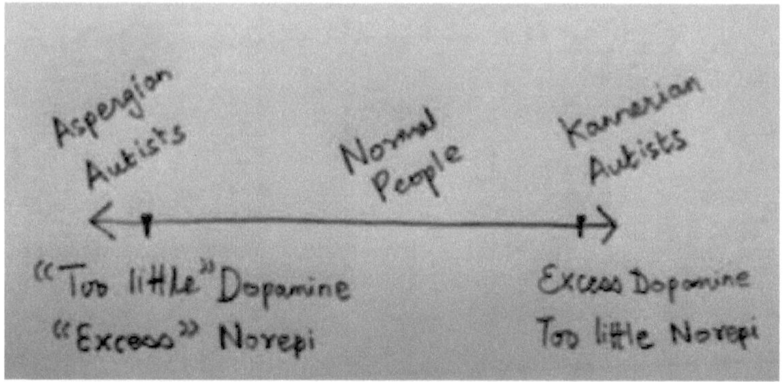

Meanings of "Aspergian"/Aspergian Autist" and "Kannerian Autist":

1. "Aspergian"/"Aspergian Autist", in the narrow sense of the term = the hyper-Norepic Aspergian/Aspergian Autist.

Cooijmans summarizes Asperger's definition of what was later called *Asperger's syndrome*: "It is a narrowing of one's relation to the world outside. It is a disturbance of instinct; an impairment of the ability to relate to or interact with the world in the direct, concrete, instinctive manner inborn to the majority of humans. [In Aspergians] Information is processed in the abstract and logical ways of the intellect, rather than instinctively and concretely. Communication is intellect-driven, rather than through the direct instinctive channels that bypass intellect. Many more symptoms result from or accompany the disturbance".

The Aspergian Autist's brain has more Norepic activity (elation). It has less gray matter and more white matter; D is less, and N is more. And now, N is the more advanced (*multiconditional signalling*) system, coming long after the ancient, indeed sponge-era, functionally simple (*monoconditional signalling*) DLM-System in the *planimal*. As the ancient Chinese said, if the right hand's pulse is greater than the left hand's pulse, only then it indicates serious, serious problem; a hypo-NLMS/hyper-DLMS condition defines the nasty type of internal disease.

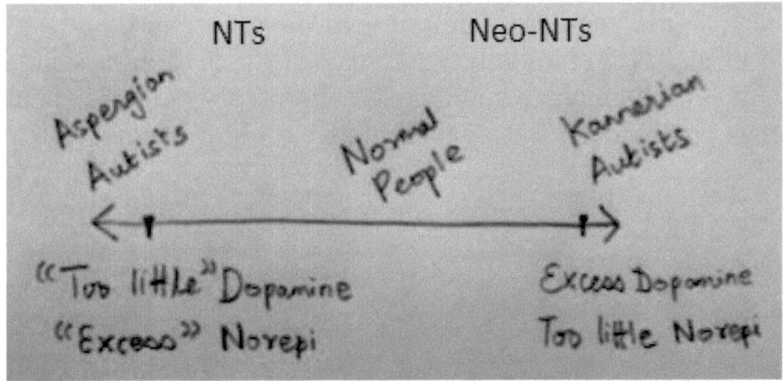

Too much happiness and elation

Serious risk of internal disease

2. The leading and powerful *"Kannerian Autist"*/Kannerian, or *hyper-dopaminergic Autist* believe themselves to be '*Aspergians*' (as being Aspergian was usually considered cooler in the Cold War era); e.g.: Temple Grandin, John Eldar Robinson (*whose great engineering remixing ability was proven in how the USSR collapsed*), Kim Peek et al. Elite Kannerians are more in parts of Finland, Russia and the US.

Kim Peek died at 56; due to extreme or divergent overclocking of their brains, Kannerians may be short lived. But middle Kannerians thrive, in forceful, competitive herds. The symptoms suffered by Kannerians: depression, Alzheimer's disease and other dementias, Ptosis, blindness, retinoblastoma (cone cancer), Insomnia, and many others.

Though he came on TV where he claimed to have 'killed a snake' – even a Kannerian 'Aspie' like Robinson may be withdrawn, isolated and otherwise culturally similar to the reclusive, hyper-logical real Aspergian.

As long as we remain as friends, it's not right to deny them an occasion to call themselves *Aspergian*, particularly given their dislike for the illogical chaos and repression hated by Norepic Aspergians as well.

We now discuss the *left* of the neural spectrum of humanity; the "Aspergians" is a group, or rather grouping of people who are hyper-Norepic (have a generally axonic, or more white-matter containing brain). This very important statement will be proven in the paragraphs below.

In babies who develop [Aspergian] Autism, white matter bundles have "more structural integrity than in normal babies", according to Wolff et al., 2012, Gao et al., 2009. Myelination is linked to White matter and Axons. It is known that glial cells have a major role in myelination.

One may think that glia are more associated with a hyper-Norepi neurostructural ideology. So a hyper-glial neural substrate is likely hyper-Norepic.

And indeed we find that Aspergian Autists are such; as John Hopkins Dept. of Neurology says: "Marked microglial activation observed in the cortical regions and white matter of [Aspergian] Autists".

The right brain controls the left body and the left brain controls the right body (*which is why it's the right pulse being stronger that is the problem*); and we see that the right brain is typically dominant in Aspergians:

C. Gillberg, 1983, found that "62% of the autistic children were left-handed compared with 37% of the controls", a finding "replicated by others". Thus Aspergians ("Autists") are likely to be right-brained, left-bodied.

Of course there are higher levels of Norepi in the right brain, and higher levels of Dopamine in the left brain, which shows again how Aspergians are more Norepic. "There is more white matter (reflecting axons) in the right brain and more Gray matter (cell bodies) in the left brain".

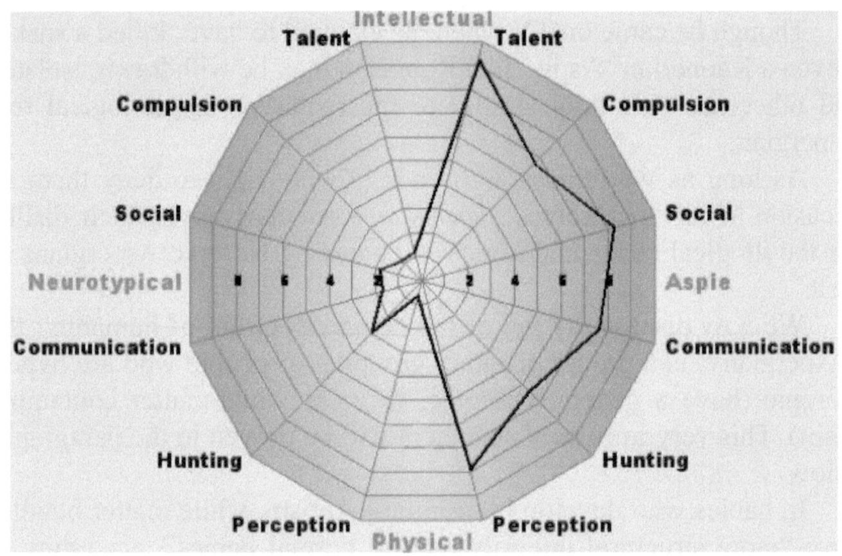

(Above: Author's "brain-map" from Leif Ekblad's old Rdos test. Below: your typical Aspergian "Droogie's" profile, the Myers-Brigg test)

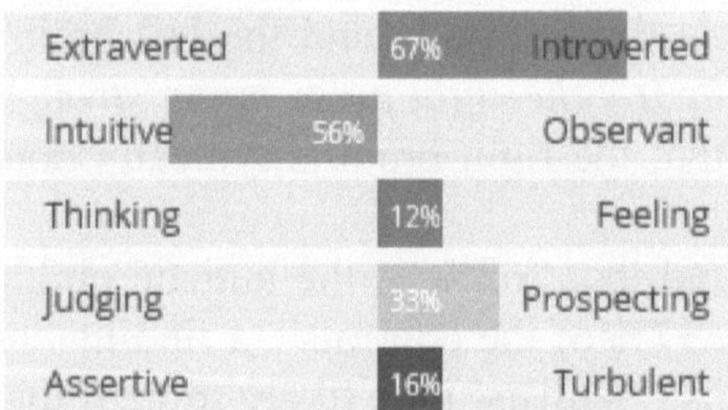

Of course the Myers-Brigg test is not always objective. The "thinking vs. feeling" idea, for example, is rather subjective.

The dominance of the right brain in Aspergians is a universal given; e.g.: "In the autism group, significant functional connectivity clusters were found in the right amygdala" (Kleinhans et al., 2008); "normal people had significantly increased connectivity in the left amygdala compared to Autists", showing how Autists are right brain-dominant.

Another proof of the fact that (Aspergian) Autists are hyper-Norepic/right brain-dominant, is how Turner et al., 2006, found less cabling out of the basal strongholds of Dopamine: "Atypically diffuse functional connectivity between caudate nuclei and cerebral cortex in autism".

Clearly (Aspergian) Autists are generally hyper-Norepic or hypo-Dopaminergic. We can say Aspergian Autists have a very "nonlinear" thinking.

Extreme syllogicians can be called "Kannerian Autists", like Kim Peek. They often find success in science or art, or in the film industry, as they are syllogically creative. They can be writers of fiction as well as empericist academics; Paul Andrews et al. in *Measuring the bright side of being blue*, show how heavy analytical thinking is linked to depression.

On the other hand, as Hans Asperger said, "A touch of Autism is necessary for success in art or science," realm of the logically creative *Aspergians*.

The term Autist, of root *autos* ("acting from or occurring within"), indicates self-generating ideas in science and art – a hyper-*NLT area* (or otherwise hyper-Norepi) type – more "intuitive", less "observant" (Kannerian Autists, however, are usually more *observant* in terms of quantity; Aspergians are usually observant of what is logical); shows how the classical "Autist" was the Aspergian type of Autist.

Where we fail is in that we, as "zombies", have no *quantitative reactions to stimuli*, which, though it is true that it is associated with depression and dementia, is made compulsory and deemed "necessary for survival", in today's capitalist world by the syllogicians aka Neo-NTs.

8: Linear and Nonlinear Thinking

Though it may seem that there is only one "common mental process", we have always seen (in the accounts of Aspergians and psychologists) the idea that there is firstly a *logical mental process* and secondly a different associative process which works with beliefs

There is certainly something called nonlinear thinking or communication, and it certainly differs from linear thinking. The simplest proof of that is **Nonverbal Learning Disability** (NLD) – overdevelopment of one's basic associative skill (linear (verbal) thinking) at the expense of his logical talent (nonlinear thinking), which phenomena implies that these different entities exist as different. An example of NLD is seen in the famous Kannerian Kim Peek. The total inability to investigate the world's mysteries except with literary guides, is the meaning of nonlinear deficiency (NLD).

Symptoms of Nonverbal Learning Disability:

"Talking too much and talking too fast".

"Deficits in coordination, problem-solving, and understanding of humour"

"Highly developed verbal rote memory"

"People with NLD have strong verbal skills and often rely on verbal communication as their main or sole method of gathering information".

Inability in [post-syllogical] mathematics e.g.: confusion of X- and Y-axis.

Underestimating the speed of cars in crossings (are generally accident-prone).

Need repeated directions or accentuation of "main points" to "understand".

Nonverbal Learning Disability, like "Borderline Intellectual Functioning" (just above mental retardation), is "extremely difficult to diagnose"... as the patient "deceptively appears to be a genius[9]".

NLD is often brushed aside, especially due to the façade of normalcy or even genius that is linked to this condition of being deficient in elation, and impotent to the non-verbal sense of any complexities.

So we see, in NLD patients, excellent syllogical talent, coupled with a miserably deprived logical talent. Clearly, this proves the existence of linear (syllogical) thinking and nonlinear (logical) thinking[10].

As J. Evans of Plymouth notes, research shows a first system which works with beliefs, and a second system that works with logic:

Evans, 1983: "Theories positing dual cognitive systems have become popular in cognitive and social psychology. Although these theories have a lot of common features, inspection of the literature reveals a number of difficult and unresolved theoretical issues".

Evans continues, "The paradigm case in dual-process theories of reasoning is belief bias [a tendency to prefer *basic association* of beliefs over the logical process]... In one study (Evans et al., 1983; Klauer et al., 2000), participants were given arguments to evaluate, whose conclusions either follow or do not follow logically, and are either consistent or inconsistent with beliefs. Research repeatedly shows that both logic and belief significantly affect decisions made, but the two seem to be in conflict within individuals".

Noting that these ill-defined neural systems are often described as "System 1" (the belief-bias i.e. syllogical system) and "System 2" (the logical system), Evans adds: "While belief-bias

---

[9] Indeed we read, in the *Economic Times*, for example: "With an IQ of 162, a 12-year-old schoolgirl in UK, Olivia Manning, has been rated brainier than Albert Einstein and Stephen Hawking. She barely needs a script – she learned her lines for a Macbeth production within 24 hours"

[10] We call it "syllogical process"; why? "All Aristotle's logic revolves around one notion: the deduction (sullogismos)"... "A sullogismos is speech in which, certain things having been supposed... something new, and different from that supposed, results of necessity because of their being so". *Because of their being so* – belief – confirms the term "syllogic"

processes do have the characteristics typically ascribed to System 1 ("fast," "parallel," "automatic"), they are certainly not 'ancient'. Nor are they shared with animals that lack a belief system. While some biases may be attributed to mismatch between the function of old cognitive systems and the current world's much changed environment (Stanovich, 2004), we need a different account for others".

System 1's belief-based, *Syllogical Associative* Mental Process; examples:

All men are mortal. Socrates is a man. Socrates is mortal. Celarent	No reptiles have fur. All snakes are reptiles. No snakes have fur.	All kittens are playful. Some pets are kittens. Some pets are playful.

A syllogical statement may be of an illusory validity only ("literal logic").

System 2 has a different Axonic Associative Mental Process; examples:

input: Smoke is visible

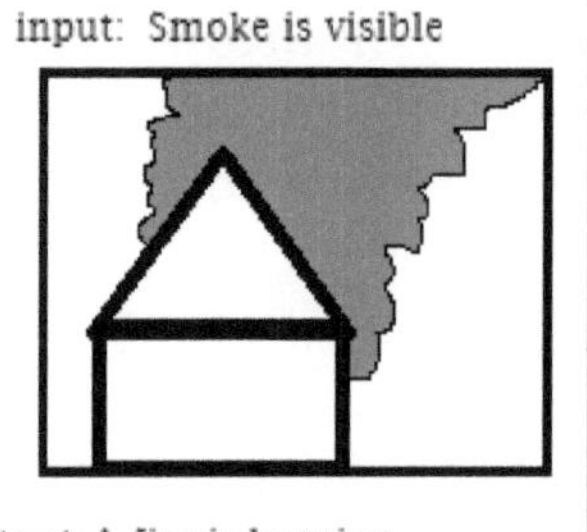

output: A fire is burning behind or in that house

input: Tree is bending

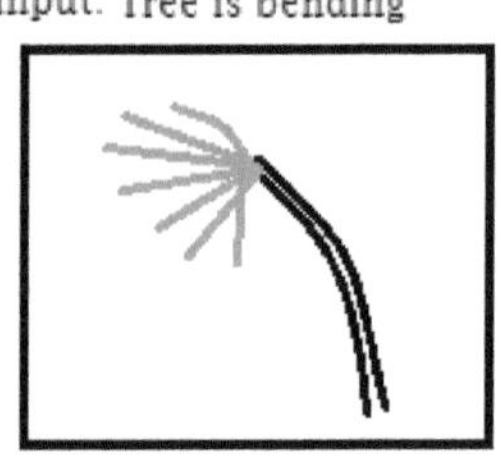

output: A strong wind is blowing

Of course verbal "thumb-rules" can compensate to an extent for the lack of such insight – a syllogician can say, "*if a tree bends to a side, the wind is blowing towards that side*" etc. and use it as a general rule; yet that policy has its grave limitations.

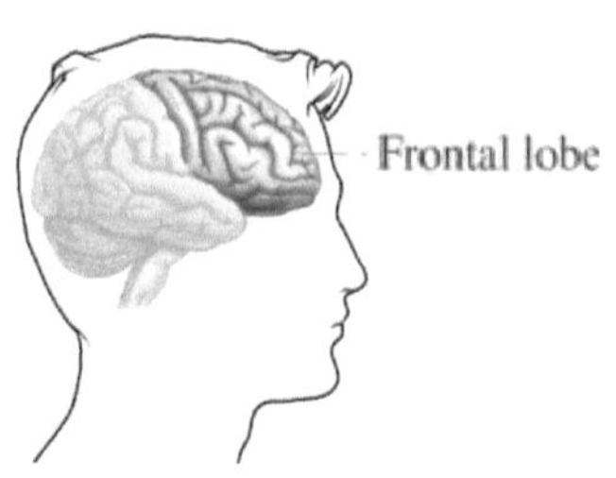

The Frontal Executive Network (FEN) of the human brain has long been recognized as having a significant role in **associative thinking**. The FEN "pulls relevant data" from posterior brain areas for the purpose of associative thinking – this much is well known.

The FEN has been recognized as the most important seat of cognition: "The frontal lobes are more developed in humans than other animals."

The FEN is key if viewing the brain in the "thinking systems" frame of reference; it can be called the *psychological nucleus of the brain.*

"There is no other part of the brain where lesions can cause such a wide variety of symptoms" (Kolb et al., 1990). J. D. Chick et al. say: "The importance of the frontal lobes derives from rich connections with almost all other parts of the central nervous system."

The ramifications of the division of the FEN into two different areas, Area 1 and Area 2, were reviewed. Also, apparently, Area 1 is relatively "dendritic", Area 2 is relatively "Axonic". Using this and other clues such as the Axon's deep relationship with Norepi; Norepi's novel datum-sensitive nature; and area 2's comparability with the olden cerebellum, etc. – a preliminary theory is proposed about how quasi-thinking takes place in Area 1 and Area 2. It is proposed that Area 1 and Area 2 handle linear and non-linear thinking (terms are clarified in this thesis) respectively. Finding these partially solves the mystery of the "*System 1 and System 2*" associated with current ***Dual system theories of cognition***, which say that **there are two different cognitive systems** in human brains.

Rabbitt et al., 1997: "Attempts to define executive function encounter an identical difficulty: no single exemplary task or even subset of tasks provides an adequate ostensive definition. It is often necessary to fall back on consensus definitions drawn from the common sense of the "man in the street" or the collective wisdom of "distinguished experts in the field"...these tend to be wide-ranging catalogues of examples of intelligent behaviour and avoid entirely discussions of underlying process". Here we dare a preliminary theory about the underlying process. The first step is to recognize that the FEN (also known as the MPFC) is made up of two parts, caudal and rostral. Gilbert et al., 2007 say: "Activation peaks from studies involving *mentalizing and self-reflection* tasks were significantly caudal to those from studies involving other tasks.

Conversely, activation peaks from studies involving *multiple-task co-ordination* were significantly rostral to those from other studies".

Gilbert et al. discuss the "adjacent but clearly distinct regions of activation within MPFC related to (i) mentalizing vs. non-mentalizing condition and (ii) SO vs. SI attention [i.e., multi-object contemplation]".

That the FEN features 2 different parts is most visibly proven in Hiram Brownell et al.'s experiment. Summary of Rita Carter's summary of the experiment: 2 stories were asked. The first story tested Theory of Mind ("mentalizing"). The second tested multi-object contemplation ability. The question relating to the first story was: Why did the burglar give up? – Which required one to think about the burglar's mind (a type of linear thinking); the question relating to the second story was: why did the alarm go off? – It required multi-object contemplation (a type of nonlinear thinking).

Paraphrasing Rita Carter: "Answering the first question required ideas regarding the burglar's mind... The scans showed that normal persons used quite separate regions of their brain to work out each answer. The first question, which called for calculation of a person's mental state, used a spot in the middle of the prefrontal cortex – one of the most 'evolved' parts of the brain" [marked orange, 1, "LT-area"]. Working out the question related to the second story used another spot beneath it" (2, yellow, "NLT-area", a spot preferentially used by Aspergians acc. to Carter).

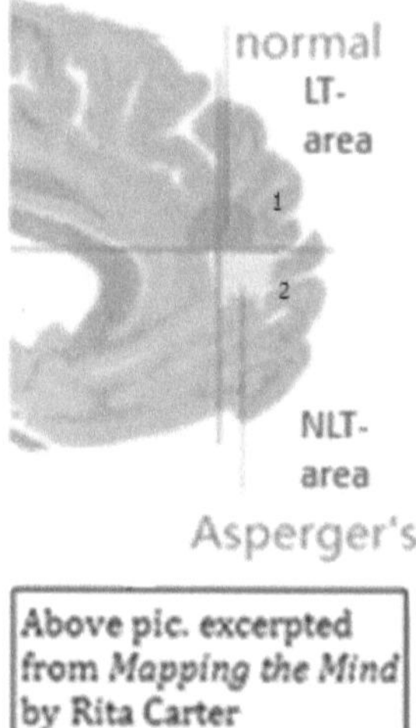

Above pic. excerpted from *Mapping the Mind* by Rita Carter

Apparently, this division is nothing but the one previously shown. I call the mentalizing-linked part, the *Linear-Thinking area* (LT-area); and the lower part, the *Nonlinear-Thinking area* (NLT-area).

And now we will study the reasons behind this nomenclature we follow.

The MPFC, or to use the other current term, "BA10" – "has been described as "one of the least well-understood regions of the brain".

"During human evolution, this area expanded relative to the rest of the brain". (Expansion of both NLT-area and LT-area took place).

Note the "dendritic" attribute of the LT-area: "It [though MPFC is said over here, more precisely the LT-area is meant] is unusual in how its neurons have particularly extensive dendrital arborisation".

"Area 10 [LT-area] in humans has the lowest neuron density among primate brains".

Thus the neurons of the LT-area, have much more dendrital development; therefore, the density of neurons or axonic fibre is low.

If the LT-area is so descried, it follows that the NLT-area, in contrast, has higher axon/neuron density, with lesser dendrital arborisation. We can represent our understanding as a rough diagram:

Figure 9: The LT Area vs. the NLT-Area

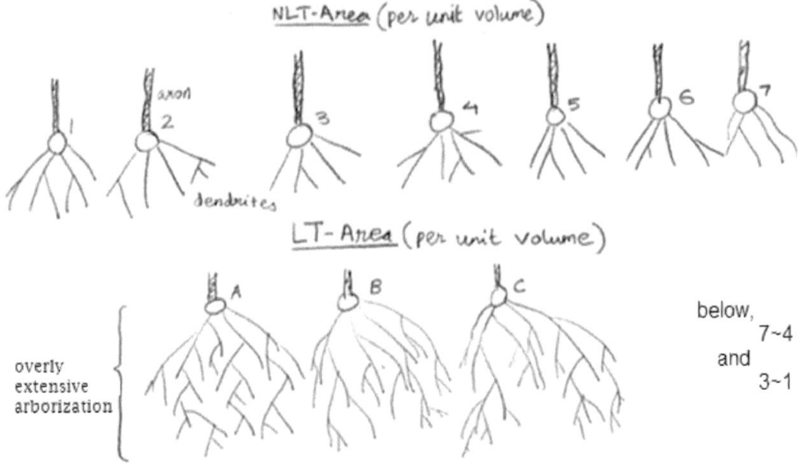

In the context of above diagram, the diagrammatic statement of the *Theory of 2 types of thinking* (linear and nonlinear thinking), is:

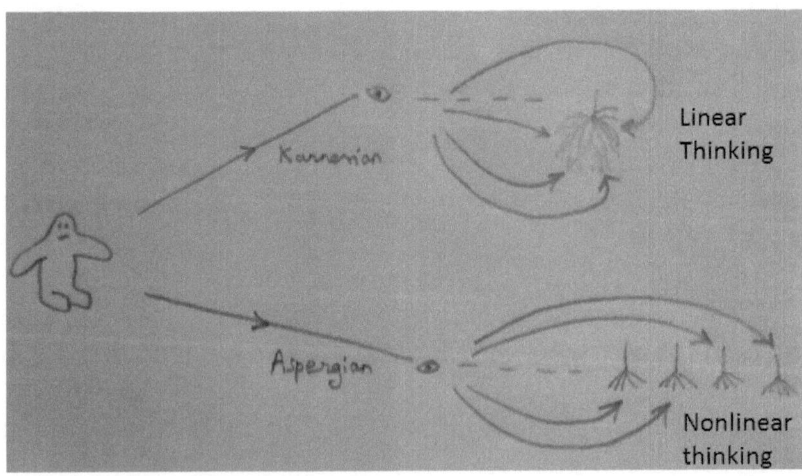

In both parts of the FEN, the neurons are emulating reality – they are storing **models of objects**, for purpose of associative thinking.

Models stored are not *reality* – they are "representations" of reality. Associative thinking presumably rejects models which do not fit the kernel of apparent truths, i.e., "the mind's worldview". Before this process, the system does not know which models are correct.

It is proposed that, due to more axons or more neurons per unit volume in the NLT-area, the NLT-area system emphasizes contemplation of many models, an ability to store more models rather than more detailed models. The opposite is the case in the LT-area. As dendrites are associated with Dopamine; Axons with Norepi – therefore, relative to the dendritic LT-area – the axonic NLT-area is clearly associated more with Norepic processing than Dopaminic processing. Let us now study this Norepic aspect of the NLT-area.

Norepic processing has a special affinity for the "odd one out" datum:

a. The P300, the Norepi-related cortical event, "responds to environmental stimuli having behaviourally relevant, motivational, or attention grabbing properties" (Johnson et al., 1993; Pineda et al., 1989; Swick et al., 1994) relative to other stimuli.

b. "The P300 is usually elicited using the oddball paradigm, in which low-probability target items are mixed with high-probability non-target (or "standard") items" (Swick et al., 1994).

c. Devauges et al., 1990, found "[Norepi] release can increase the alteration detection rate (number of times an alteration was selected) in multiple-cue probability learning".

d. A. J. Yu et al., 2005, developed a Bayesian framework to examine NE release in instances of "unexpected uncertainty," where alteration in sensory information produces a large disparity between expectations and what actually occurs. The model predicts that NE levels spike when the predictive context is switched, then subside. It was shown that lesions of the LC impair this attentional shift".

The above clues suggest that Norepic control, which occurs far more in the NLT-area than in the LT-area, has a special affinity for the "odd one out" datum; so does the model of axonic processing.

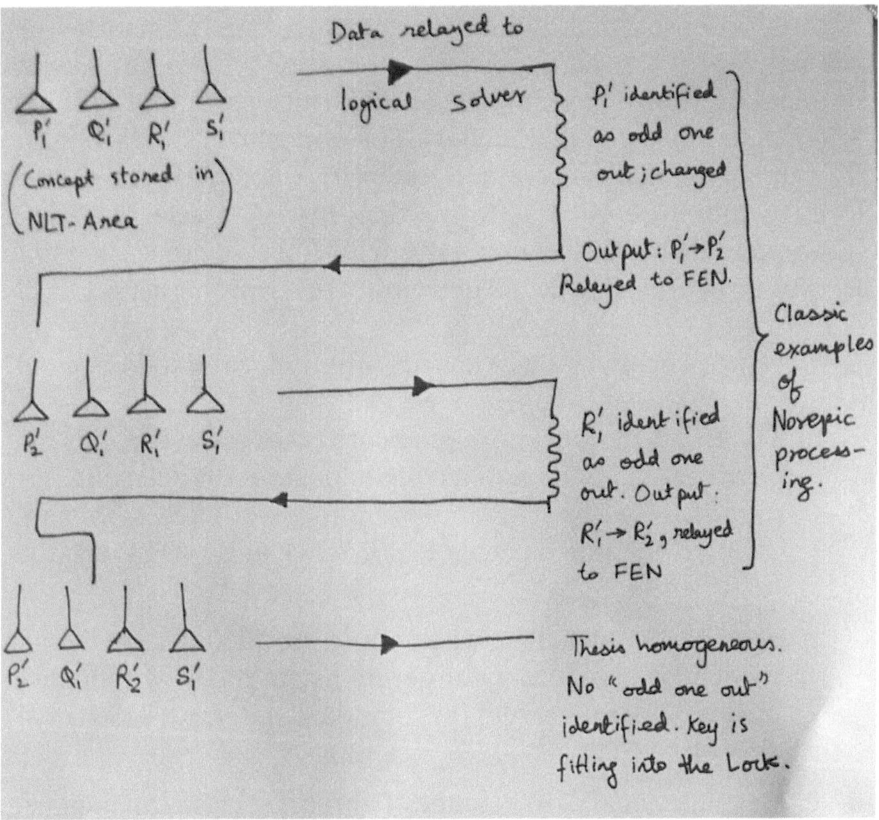

The (mind's) eye-catching datum becomes, then, the subject of modification or expansion in a multi-threaded thought process – this is how axonal processing works (oft in the relatively axonic NLT-area).

Now we will explore how dopaminic processing, which can be seen more in the LT-area-linked dendritic circuit, differs from the Norepic.

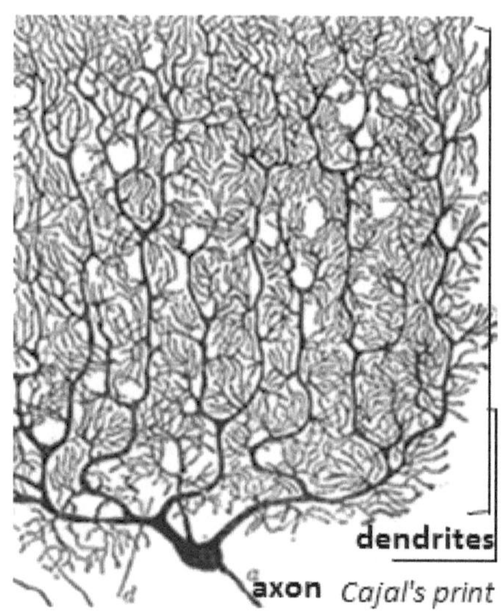

Cajal's print

Above shown are the cerebellum's Purkinje Cells. They are very dendritic. In that, the cerebellum resembles the (meta) LT-area, whose unique dendritic nature we had discussed in the last few paras.

Extensive dendrital arborisation is not very common, being found in only a few areas like cerebellums and LT-areas. In apes, cerebellumicity is less in the Frontal lobes. Man is uniquely the creature in which cerebellumicity has cropped up in the frontal lobe, leading to Huxley's quote: "You never see wild animals going through the absurd and often horrible fooleries of magic and religion; dogs do not ritually urinate in the hope of persuading heaven to do the same and send down rain. Asses do not bray a liturgy to cloudless skies. Nor do cats attempt, by abstinence from cat's meat, to wheedle the feline spirits into benevolence. Only man behaves with such gratuitous folly. It is the price he has to pay for being intelligent but not, as yet, quite intelligent enough." It is the difference between the life of the cat and the rat; studying the ancient cerebellum, gives clues as to how dendritic processing works.

If we were to put this in a summary: we can theorize that the meta-LT-area's style of processing can be understood by an examination of the processing style of the similarly dendritic cerebellum.

The cerebellum "receives input from sensory systems of the spinal cord and other brain areas, and integrates these inputs to fine-tune motor activity". E.g.: "ensuring you don't step on your right foot with the left"; hmmm; thus "a standard test of cerebellar function is to reach with the fingertip for a target: The Healthy person will move the finger in a rapid straight trajectory, while a person with cerebellar damage reaches slowly and erratically, with many mid-course corrections". So we'll say the cerebellum's output (*where limbs should be*) is a set of short-range logical operations performed on inputs (e.g.: *where limbs are*, and other data).

What is the similarly dendritic LT'area's Dopamine doing? It takes data, and, with the help of the PFC, does a pre-programmed (rote-based) modification or rearrangement or trimming of that data (which can be, for example, an IF-THEN type operation, often seen in them).

The theory that the *dendrital/dopaminic associative process* may involve *rearrangement* of data, is again upheld: "Aoccdrnig to rscheearch at an Elingsh uinervtisy (16), it deosn't mttaer in waht oredr the ltteers in a wrod are, olny taht the frist and lsat ltteres are at the rghit pcleas. The rset can be a toatl mses and you can sitll raed it wouthit a porbelm." What actually occurs when we look at, say, "lteter?" Surely we can say only that – (in a dendrital process) the input data was compared with innate data (memory of "letter"), resulting in an output (mental contemplation of "letter"); it was, thus, a *rearrangement* of lteter. This is not so different from what happens in the cerebellum, if we think about it!

A modification or rearrangement or trimming of data – short-ranged logical operations – is seen also in the similarly dendritic, therefore more Dopamine-based cerebellum – it "receives input from sensory systems of the spinal cord and other brain areas, and integrates these inputs to fine-tune motor activity". E.g.: "ensuring you don't step on your right foot with the left". The cerebellum's output (*where limbs should be*) is a set of short-range logical operations performed on inputs (e.g.: *where limbs are*, and other data).

(And what makes us so sure that the LT-area is involved in the above example? The fact that Aspergians are bad at jumbled words!)

Surely, in contrast to the NLT-area that is more full of Norepi – the LT-area, being as Dopaminic and dendritic as the cerebellum – is undertaking a lot of short-range logical operations. There is quantitative booster gate signalling in the brain, making them adept at operations such as filling forms and getting by in the Theater of the Absurd, by going to every dim alley and mugging all the relevant details, but unrelated to their often-weak logical abilities.

If **Norepic association** involves weighing of a datum within a handy context, **Dopaminic association** involves quasi-syllogical data modification with respect to pre-existing data or pre-defined formats.

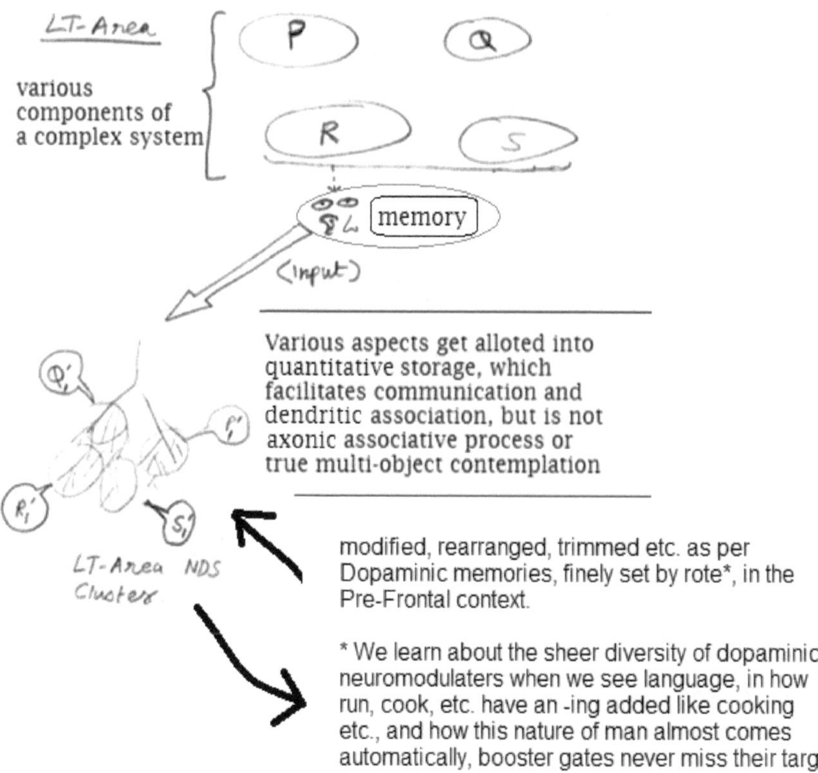

Thus, one may theorize that in cerebellum-like, or dendritic neurosystems - circuits with high quantity of data (where is foot, what is terrain like etc.) carry out an associative process which takes into account many sensorily intaken as well as non-intaken, innate data - but it involves lesser *deduction hinged upon eye-catching data*, in the style of Norepic processing. Therefore, as the example of cerebellum shows, the output of the dendritic system (e.g.: LT-area) is about homogenizing or rearranging input data into a pre-defined coherence, not *odd-one-out*-eliminating, logical output. Examples of dendritic processing, which most likely uses the LT-area:

1) How routine math is solved.
2) How ideas are reduced to outputs which are coherent with respect to Givens ("knowns" and "seens") - even if not necessarily logical. This type of mental activity is called syllogic.

Dopaminic processing, modification and rearrangement of input data - occurs a lot in games like crossword, chess, jumbled words, and so forth (and especially the latest game of Sudoku); that is why Aspergians don't like such games or why these games occur in the "normal" newspapers printed by syllogician power elites.

An example of complex dopaminic processing hinging on the LT-area: Given the many details one knows regarding many people - "what to do"? All these details are arranged for a '*do this*' result (e.g. intimidate, appease etc.). Such are the "situations for which there are no well-rehearsed or well-specified ways of behaving and therefore behavioural organization needs to be self-determined".

Reminds us of the Aspergian's uneasiness with situations demanding quantitative mental activity e.g.: interacting with The Absurd (an huge amount of quantitative thinking!). But some are good at it.

E.g.: In the documentary "*Street Thief*" the thief has "expertise in social engineering, stalking, and intelligence gathering"; the thief "discusses his careful, meticulous planning cycles". The PFC's data-rearranging neurostructures, which one might call "QRR modulators" - and which often correspond to "instincts" -

constitute a "primal logic" elsewhere defined as RQ (Receptive Quotient), which, though often engaging in complex calculations, involve more Dopaminic or booster gate calculators than Norepic calculators.

While good at pertinent, qualitative conversation, Aspergians are, for example, weak at quantitative conversation ("small talk"). Burgess notes that impairment of the LT-area "led to a tardiness and disorganization [*Aspergian/right brainer traits*], the severity of which ensured that despite his intact intellect and social skills [some of the higher social skills seem dependent on the NLT-area], the patient never managed to return to work like before". All this explains why Aspergians are worse at chaotic quantitative empirical activities (e.g.: office algorithms) or why they care more about reform (thus unnecessary things need not be quantitatively done).

Now the right brain is relatively axonic/Norepic so the logical system should involve more the right hemisphere... And indeed Bowden et al., 2003, note that the right hemisphere is associated with the "*Aha! Insight experience*", which is nothing but the logical process.

In summary, we can say:

1. The NLT-area is about several models occupying separate neurons

2. The LT-area is about fewer but more elaborate/detailed models; its dendritic organization reflects a structural ideology conducive to communication, multitasking, and association of beliefs/"Givens" (e.g.: solving Rubik's cube).

The former is the scientific/logical method - because in the philosophy of science, under any given subject, considering as many models as possible is most scientifically sound (Lakatos et al., 1978).

9: Linear Thinking and Receptive Quotient

Is there a link between what we can call **receptive social activity,** and a hyper-Dopaminic mentality? "Socially stimulated" fruit flies have three times the amount of dopamine in their brains than "socially deprived" fruit flies" (Indrani et al., 2006). Similarly:

"Early Social Experience Is Critical for the Development of Dopamine Modulation of Prefrontal Cortex Function (Petra et al.)".

> And:
>
> "Increased social status and support correlated with the density of dopamine D2/D3 receptors" (Martinez et al.).

We'll see that a lifestyle or ideology of *quantitative real-time (especially verbal) interaction* is linked to a heavily-booster gate-based neurostructural ideology. This along with other "largely booster-based" intelligences, may be called *Receptive Quotient*, or "RQ".

Logic-based lifestyles should be explored; they are qualitative (involving qualitative rather than quantitative signalling), partly because they are logically optimized, hence involve less mental processing.

By eating a lot, the Neo-NTs may go with the ordained "deep memorization" routine, but, as the example of Peek dying in his 50s shows, it is generally harmful to mental health (that does well with maintaining as low a metabolic rate as possible; see chapter 10).

A general law to identify RQ-type circuits is that they are quantitatively dependent on external stimuli (receptive). Example of a RQ circuit -- in *Obsessive Compulsive Disorder* (OCD), patient responds to a large quantity of dirt, which is the "*quantitative response to quantitative caches of external stimuli*" which one must look for in figuring out whether or not more DLMSs are here involved.

If people react to a large quantity of superficialities; *having a quantitative mode of function*) instead of dealing with a few core matters (*a qualitative mode of function*) – "quantitative responses to quantitative caches of external stimuli" – that is what we look of, if able to say, "Oh, this is linear thinking or processing", that is, RQ.

Your usual serial killer, before finalizing his target, checks many things – whether or not anyone is nearby, whether or not all doors are closed, whether a road is nearby, a spare set of clothes is available: an external stimuli-oriented function involving more DLMSs.

Verbal (quantitative) theory of mind registers, quantitatively enough, others' minds' contents (external stimuli), for later use in solution-fetching. This, a widespread practice in the eastern and western worlds, is a heavily RQ-type practice, and is perhaps the single biggest culprit that causes dementias throughout Finland to USA.

Though called "verbal theory of mind", it is different from, for example, the Empath's qualitative Theory of Mind. A number of novel social talents are seen in, for instance, the Negroes or other autochthons. These are typically Norepi-involving right brain processes (Brownell et al., 2000), so are not generally hyper-Dopaminic, and are thus not classified as RQ even though they are social.

Solving the above joke involves some verbal calisthenics that Aspergians seem rather powerless to perform, but NLDs eta al. easily do.

A solution-fetchist lifestyle is described in the 48 laws of power:

1. Never outshine the master.
2. Never put too much trust in friends, learn how to use enemies.
3. Conceal your intentions.
4. Always say less than necessary.
5. So Much Depends on Reputation – Guard it with your Life.
6. Court attention at all costs.
7. Get others to do the work for you, but always take the credit.
8. Make other people come to you; use bait if necessary.
9. Win through your actions, never through argument.
10. Infection: avoid the unhappy and unlucky.
11. Learn to keep people dependent on you.
12. Use selective honesty and generosity to disarm your victim.
13. When asking for help, appeal to people's self-interests, never to their mercy or gratitude.
14. Pose as a friend, work as a spy.
15. Crush your enemy totally.
16. Use absence to increase respect and honor.
17. Keep others in suspended terror: cultivate an air of unpredictability.
18. Do not build fortresses to protect yourself. Isolation is dangerous.
19. Know who you're dealing with; do not offend the wrong person.
20. Do not commit to anyone.
21. Play a sucker to catch a sucker: play dumber than your mark.
22. Use the surrender tactic: transform weakness into power.
23. Concentrate your forces.
24. Play the perfect courtier.
25. Re-create yourself.
26. Keep your hands clean.
27. Play on people's need to believe to create a cultlike following.
28. Enter action with boldness.
29. Plan all the way to the end.
30. Make your accomplishments seem effortless.
31. Control the options: get others to play with the cards you deal.
32. Play to people's fantasies.
33. Discover each man's thumbscrew.
34. Be royal in your fashion: act like a king to be treated like one.
35. Master the art of timing.
36. Disdain things you cannot have: ignoring them is the best revenge.
37. Create compelling spectacles.
38. Think as you like but behave like others.
39. Stir up waters to catch fish.
40. Despise the free lunch.
41. Avoid stepping into a great man's shoes.
42. Strike the shepherd and the sheep will scatter.
43. Work on the hearts and minds of others.
44. Disarm and infuriate with the mirror effect.
45. Preach the need for change, but never reform too much at once.
46. Never appear too perfect.
47. Do not go past the mark you aimed for; in victory, learn when to stop.
48. Assume formlessness.[11]

10: The Root Cause of Dementia

Is an overly dendritic neural doctrine, as seen in the LT-area, in any sense problematic? "Aristotle and Galen did not consider the cerebellum part of the brain: They called it the parencephalon ("same-as-brain"), as opposed to the encephalon or brain proper". Bower argues that the cerebellum is a sensory organ" – hyper-receptive, unlike the rest of the brain, which is a bastion of Norepi.

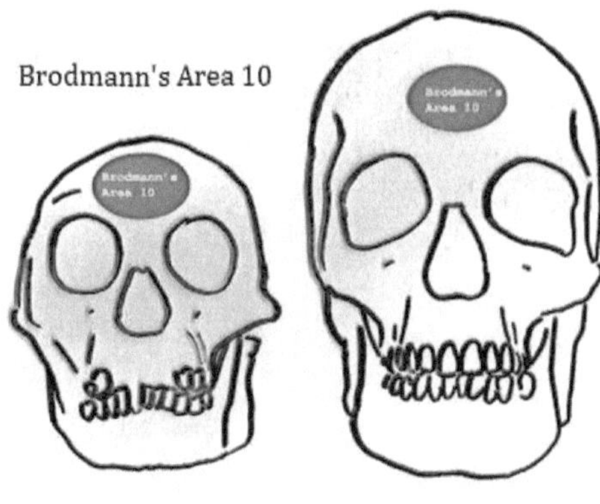

homo floriensis *sample modern man*

Is the LT-area/FEN's "evolution" towards a cerebellum-like dendritic structure – a "regression"; is it a problem rather than a progress? It might seem a medical emergency given Fingelkurts' finding that depressed person's brain is less specialized (more gray matter type, more dendritic). We can study the extinct *homo floresiensis* (HF) to answer that question. Though the LT-area did indeed expand as man evolved – it is notable that, in the HF, BA10 (LT-area) is *even larger than it is in modern man*. Notably, it was accompanied by progressive brain shrinkage (microcephaly) – though HF was advanced (a derivative of *Homo erectus*), its brain became smaller than the chimpanzee's. Indeed the HF became extinct, eventually... question is, and did the original Atlanteans really cause them to go extinct? HF became extinct at least partly due to its fruitless bid to retire to the monkey's life in a world of men. For the innocence of that reason, HF-derived people are also innocent.

The *rise and fall of the HF* is a closed model which proves that *increased reliance on the LT-area may be the wrong way to use the brain*, as it causes shrinkage of the brain over time. The more important question - can we say so generally? Is *increased use of the LT-area and the other related, largely Dopamine-oriented brain areas* - a general risk, causing many other psychosomatic health problems?

In this chapter, scariest of the chapters written until now, we explore this mysterious question, the question of a dreadful answer.

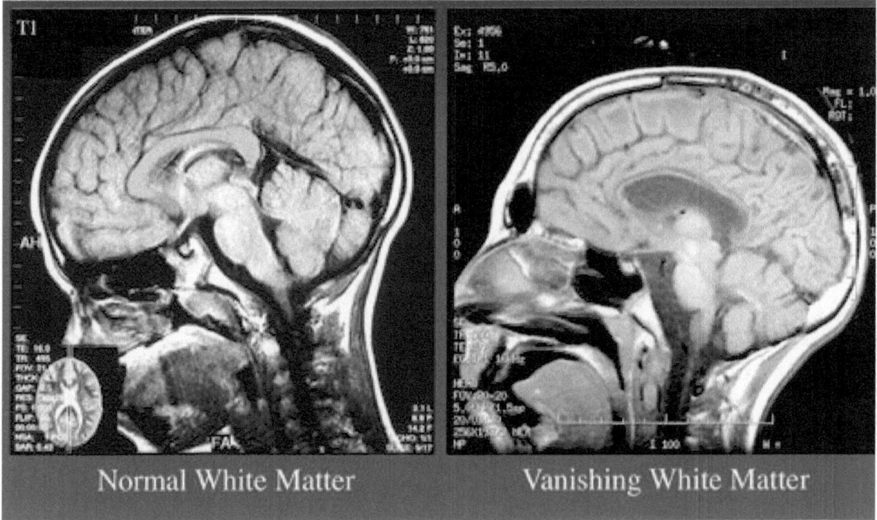

Normal White Matter | Vanishing White Matter

Dopamine can almost be called "alien" in the (upper) brain, an organ that originally was the home turf of Octopamine (and later, Norepi) ever since the evolution of brain in the animant, as we've seen.

Proofs are the following statements:

- Dopamine can be called alien in the brain because it is rare: "The prefrontal cortex is one of the very few cortical areas to receive a dopaminergic innervation" (Bruin et al., 2011).
- Dopamine can be called alien in the brain since evidently the upper brain is the *natural ecology for Norepi* – it can be plainly inferred from how *Norepi* travels farther (as seen below), has a greater range of operation across the brain –

as opposed to Dopamine (left), whose regions of operation are limited:

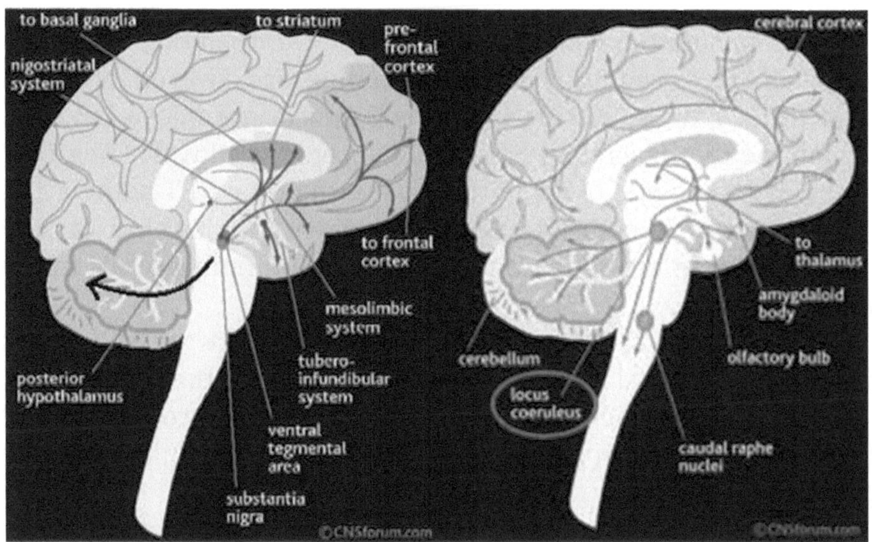

Nearly half the body's dopamine is outside the brain, there governing linear processes which need only its monoconditional signalling.

Increased presence of Dopamine in some regions of the brain is "anomalous", e.g.: the prefrontal cortex, which Morón et al. describe as a "region with low levels of the dopamine transporter" (DAT).

Morón et al., 2002, say, in the striatum and basal ganglia, D is "inactivated by reuptake via the DAT. In the **prefrontal cortex**, however, there are very few DAT proteins, and **dopamine** is inactivated instead by reuptake via the Norepinephrine transporter".

We notice above that Dopamine, in the prefrontal cortex, is enjoying quasi-calculative roles (as part of complex RQ circuits) because of Dopamine having taken over complex Norepi-style substrates.

Thus like the exploiter of the cuckoo, the Dopamine in the **prefrontal cortex** is enjoying the NET, the Norepi transporter. It shows the meaning of saying that *Dopamine is alien in the upper brain.*

(PFC and higher cortical layers is the upper brain, while the neural tissue area located around the caudate nucleus is the lower brain).

The calculative benefits availed to PFC-Dopamine, which are not available to caudate nucleus Dopamine, are covered by Yavich et al. (2007), who describe a high-tech *slow decay* enjoyed by such Dopamine that uses the PFC's NET (Norepic) pathway: "The DAT pathway is an order of magnitude faster than the NET pathway: in mice, dopamine concentrations decay with a half-life of 200 milliseconds in the caudate nucleus vs. 2,000 milliseconds in the PFC."

Thus though the lower brain (Ganglia, caudate nucleus etc.) is Dopamine's natural home, we cannot say that for the upper brain, i.e., PFC and its backyard, where the increased proliferation of Dopamine, as seen in people with high RQ, may be linked to some problems.

That is, problems may arise due to general hyperactivity of Dopamine in the upper brain, as will be discussed in our theory of dementia.

Certainly, increased activity of Dopamine in some brain areas, and/or decreased activity of Norepi (lesser grey matter and/or increased white matter), is associated with many disorders of the brain:

- More Dopamine in FEN translates to Epilepsy ("Decreased Dopamine D2/D3-Receptor Binding in Temporal Lobe Epilepsy" (Werhahn et al., 2006); (less binding = increased availability).
- More Dopamine in PFC translates to Obsessive Compulsive Disorder ("Low level of D2 receptor binding" (Denys D, 2007); less binding means more activity); or Schizophrenia ("Schizophrenia: More dopamine, more D2 receptors", say Seeman et al., 2004).
- Lack of Norepi (white matter) in one region translates to Psychopathy ("the white matter tract *uncinate fasciculus* is disrupted in psychopathic individuals", Hoppenbrouwers et al., 2013)
- Deficits in the Norepic right brain translates to Non-verbal Learning Disability ("Brain scans of individuals with NLD

often confirm mild abnormalities of the right hemisphere" (Sue Thompson, 1995) which is primarily the bastion of Norepi

- Lack of white matter in the *corpus callosum*, translates to Kim Peek's Disorder (ACC: "complete or partial absence of the corpus callosum, the band of white matter" (Dobyns, 1996).

(i) Dopamine and Norepi exist in a "Yin-Yang relationship" – that is, excessive Dopamine in the brain generally implies a lower level of Norepi, and vice versa. Due to many reasons (e.g.: the D system's expansionism i.e. competition for similar adjunct chemicals (*e.g.: how D occupies the NET pathway*) – due to the structural and functional similarity of Dopamine and Norepi) – Dopamine and Norepi exist in a seesaw-like Yin-Yang relationship (proven by the numbers given about *Norepinephrine deficiency* – chapter 7 - five to tenfold elevation of D levels in those with less Norepi).

Heneka et al., 2010 write:

> "Alzheimer's patients show 70% loss of Locus Coeruleus (LC) cells"
>
> And
>
> "Degeneration of the LC may be responsible for higher Aβ deposition."

The health of the LC, a producer of Norepi, is correlated with N levels; also for instance, we see how Lyness et al., 2003 says: "In AD, reduced brain noradrenaline levels are often reported" – therefore, Alzheimer's disease is characterized by low N levels, and hence, from **(i)**, may be associated with increased presence of Dopamine...

This shows that an overly DLMS-based, NLMS-deficient brain could be at a significantly greater risk of dementia. It appears that a *high RQ* brain in which Dopamine is excessively present in a calculative capacity, so there is more dendrital (grey matter) development – suffers damage because of greater stress levels due to higher signal density; and it is some kind of quantitative stress – it is unknown if it is a mechanical, chemical, or a thermal stress, or a combination of various stresses, causing a spectrum in dementias.

The theory of how a relatively hypo-N, hyper-D state is linked to AD – is supported by how, in AD, "brain atrophy has been proposed to be left lateralized" (Derflinger S et al., 2011) (more atrophy in the left hemisphere) – in the light of how, in the left hemisphere, D tends to dominate, while N dominates in the right hemisphere.

Thus certainly we can say that in a brain overly dependent on RQ-type circuits – there are too many DLMSs; a monoconditional, simplistic, archaic type of gate is being overused in the circuital architecture.

In the old computers that used vacuum tubes, as Tony Stockill says, "With each tube generating heat to work, the cooling problems were huge. Large blowers and cooling fans around the tubes, as well as air conditioning were standard. Liquid cooling was also used". This old technology-based system generated more heat.

As he adds: "When transistors came along, there was less heat per circuit".

This modern technology-based system generated far lesser heat. Brains can be understood by that analogy – a brain which is largely dependent on dopaminic processes experiences more internal heat than a brain largely dependent on Norepic processes.

Just how olden electronic circuits used many more parts than new ones, for similar jobs – an overly DLMSs-based neurocircuital architecture also uses an order of magnitude more DLMSs than usual.

That is why we make such a model of normal and at-risk brains:

Normal brains, in which there are say like **1n trillion** Norepi and **20n trillion** Dopamine check posts; and at-risk brains, where there are, say, a lesser than a **trillion** Norepi and **100n to 200n trillion** Dopamine check posts; since each simple booster gate (DLMS) as well as long-range booster gate (NLMS) activity adds the same amount of energy to the net signal energy, a type B brain experiences an overall greater energy density and a higher mental metabolic rate.

If the volume of D/N-*signalling* is somewhat correlated with the *number of D/N gates* present, which is likely – then normal brains engage in qualitative, at-risk brains engage in quantitative signalling.

Thus a largely DLMS-based brain experiences a greater signal density.

The increased energy/signal density (which is, evidently, due to more dopamine logic gates and quantitative computation) in case of Alzheimer's disease, is discussed by Alice Walton of Forbes:

> "People who have more activity in their default mode networks may have increased risk for Alzheimer's disease. As Holtzman of Washington University says, "people whose default mode networks have an average increase in activity relative to others may be at increased risk to get Alzheimer's disease later in life; lesser the activity in this network, lesser the risk".

We saw the *default mode network*, at whose heart is the biggest culprit, the cerebellumic LT-area, which may be behind a process in which Dopamine harnesses Norepi pathways to solve logical problems with extreme syllogical precision, if not logical correctness.

The increased signal density in Alzheimer's is confirmed in how the lateral ventricles, which carry a fluid to support neural activity by attending to metabolic waste etc., are enlarged in Alzheimer's disease; that once again indicates excessive metabolic rate.

(Interestingly, the ventricles are enlarged in bipolar disorder as well).

The "Amyloid" debris associated with AD -- is pronounced in the DPFC (Murray et al., 2012), where most RQ-related calculative tissues are yet where Dopamine action, according to my theory, is anomalous.

So what is the pathological process behind dementia? Though there may be a whole gamut of problems due to oversupply of D and undersupply of N, the main problem may be, simply put, the quantitative presence of signals, the high mental metabolic rate, which is a devastating thing for brain tissue. Given how AD brains disintegrate and shrink, it's tempting to conjecture that the damage may be thermal. How? Fact is, each time a signal passes through a signal cable, some loss occurs on the way. This loss is converted to heat (electrical equal of friction). Greater energy or signal density, as in hippocampus[11] (handles conceptualization), Entorhinal Cortex, PFC etc. may imply overheating, which may cause neuronal collapse and "neuroinflammation" (that is a repair reaction).

Effect of overheating because of excessive monoconditional signalling:

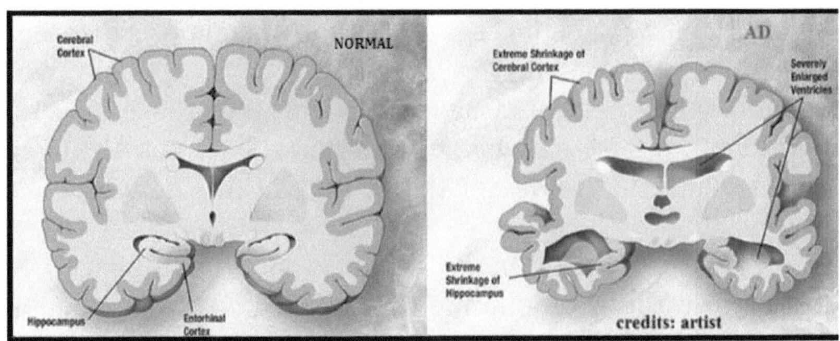

credits: artist

Heavy neural signalling seems to burden the little-explored microtubules.

Microtubules, traditionally known as involved in maintaining the cell structure, are also "required for the establishment of cell polarity".

Microtubules are "affected by signalling pathways", and this may "contribute to transmission" (Gundersen et al., 1999). "The signals had to be confined to individual tracks... the most promising candidate for this function seemed to be microtubules" (Buehler, 1985-2015).

---

[11] We read: "Dopamine modulates the hippocampus, a crucial brain system for memory". And there are "Interactions between midbrain dopamine regions and the hippocampus"... see, in a quantitative type of brain, the type of consciousness is itself is of a quantitative rather than qualitative type; the hippocampus, the brain area responsible for memory and consciousness, is, for this cause, one of the first to burn out

Microtubules are key components of the signalling process: "Microtubules are regulated, dynamic polymers that respond to signalling pathways by changing their dynamics and organization in ways that may contribute to the signal transduction process" (Gundersen, 1999). Microtubules help in signalling operations, therefore a lot of signals means a lot of microtubule activity. And indeed we find that the microtubules have disintegrated, in Alzheimer's:

Matsuyama et al., 1989: "AD is a clinicopathologic syndrome of unknown etiology with numerous abnormalities in neuronal and non-neuronal cells. A review of the literature suggests that a common basic intracellular defect may underlie many of the reported abnormalities. We hypothesize impairment of the microtubule (MT) system as one explanation for the pathogenesis of AD".

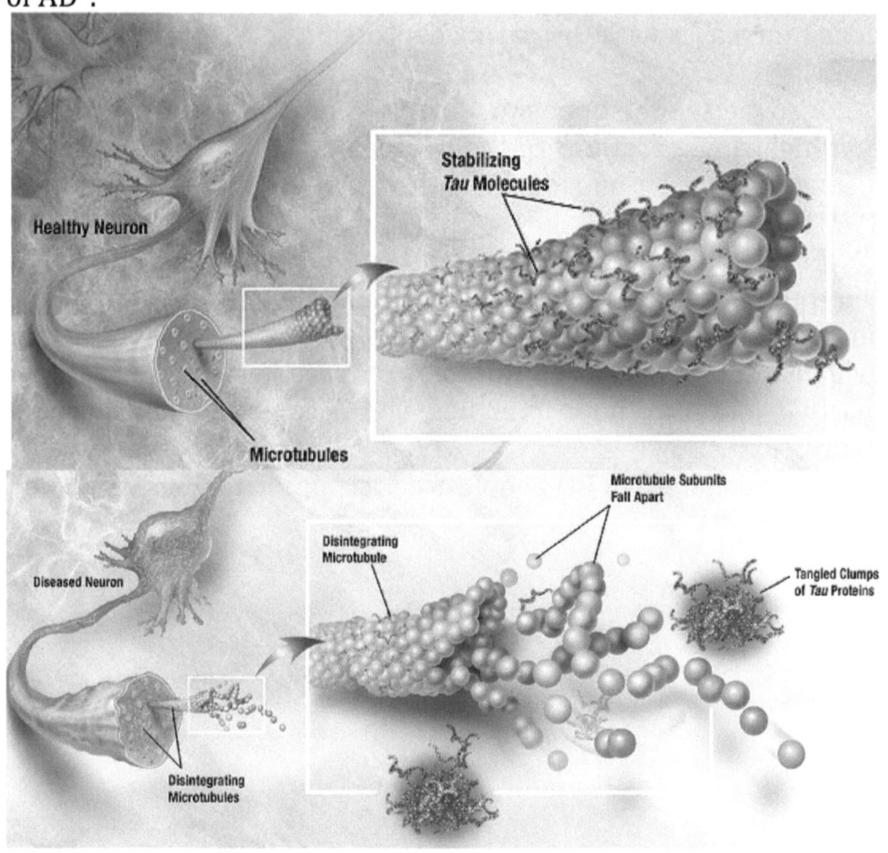

Above picture describes the breakdown of microtubules in AD.

So what destroys neurons and their constituent microtubules, in AD? A higher signal density, due to a dendritic neurostructural ideology.

Dementia is no lone wolf; this family of diseases also includes depression. This we say because similar observations are made in depressed and demented brains: "Default mode network shows greater activity when depressed participants ruminate" (Cooney et al., 2010).

Depression also involves enlargement of lateral ventricles (Kempton et al., 2011), like AD. That depression does indeed fall in this disease family is seen in how right brain issues accompany Depression: it is associated with weak *right* amygdala modulation by reappraisal on trials featuring negative pictures, and this deficit increases with increasing depression severity (Dillon et al., 2013).

Also we see the opposite of depression: "Studies have shown that high concentrations of the neurotransmitter Norepi leads to feelings of elation and euphoria [extreme happiness] (Franken, 1994)."

Further we read, from *Mental Health Daily*: "When levels [of Norepi] are low, a person can lack motivation, healthy cognition, and intellectual skills – all of which are vital for success... Other evidence shows that when Norepi is reduced in rats "they become inactive, and have poor appetites, and become model aspects of depression".

Conclusion: High RQ people are susceptible to depression and dementia.

11: Recommendations

Thus as the progression of dementia is caused by quantitative signalling in RQ type processes, which are promoted when people run into each other quantitatively, it's easy to end the epidemic of dementia.

Mankind must spread out, into a living style that goes with, not against, one of the laws of physics, *the law of ever-increasing entropy.*

12: References

- Behrendt RP, Hippocampus and consciousness, 2013
- Brownell et al., Cerebral lateralization and theory of mind, 2000
- Bowden, Edward M.; Jung-Beeman, Mark (1 September 2003). "Aha! Insight experience correlates with solution activation in the right hemisphere". Psychonomic Bulletin & Review 10 (3): 730–737. doi:10.3758/BF03196539
- Buehler, Are microtubules the 'nerves' of the cell?. Buehler, G. Albrecht, Is Cytoplasm Intelligent too? In: Muscle and Cell Motility VI (ed. J. Shay) p. 1-21
- Burgess, P. W., Veitch, E., & Costello, A. (submitted). The role of the right rostral prefrontal cortex in multitasking: The six element test.
- Carlo Zancanaro et al., Biogenic Amines in the Taste Organ, 1997
- Carter, Rita, Christopher D. Frith, Mapping the Mind
- Chapman, R.M. & Bragdon, H.R. (1964). Evoked responses to numerical and non-numerical visual stimuli while problem solving. Nature, 203, 1155-1157.
- Chen J. Y., Bottjer D. J., Oliveri P., Dornbos S. Q., Gao F., Ruffins S., Chi H., Li C. W., Davidson E. H. et al. (2004). "Small bilaterian fossils from 40 to 55 million years before the cambrian". Science 305: 218–22. doi:10.1126/science.1099213.
- Cooney, R.E.; et al. (2010). "Neural correlates of rumination in depression". Cognitive Affective and Behavioral Neuroscience 10 (4): 470–478.
- Cross AJ, Crow TJ, Perry EK, Perry RH, Blessed G, Tomlinson BE., 1981, Reduced dopamine-beta-hydroxylase activity in Alzheimer's disease.
- Daniel G. Dillon, Randy L. Buckner, Sunny Dutra & Diego A. Pizzagalli, DEPRESSION IS ASSOCIATED WITH WEAK RIGHT AMYGDALA MODULATION DURING QGAOTIONAL REAPPRAISAL, 2013

- Denys D1, van der Wee N, Janssen J, De Geus F, Westenberg HG, Low level of dopaminergic D2 receptor binding in obsessive-compulsive disorder., Biol Psychiatry. 2004 May 15;55(10):1041-5.
- Derflinger S et al. (2011), Grey-matter atrophy in Alzheimer's disease is asymmetric but not lateralized., J Alzheimers Dis. 2011;25(2):347-57. doi: 10.3233/JAD-2011-110041.
- Desikan et al., 2010, Selective Disruption of the Cerebral Neocortex in Alzheimer's Disease, the Alzheimer's Disease Neuroimaging Initiative
- Devauges V, Sara SJ, Activation of the noradrenergic system facilitates an attentional shift in the rat. Behav. Brain Res., 1990 Jun 18;39(1):19–28.
- Dewey et al., 1995, Serotonergic Modulation of Striatal Dopamine Measured with Positron Emission Tomography (PET) and in viva Microdialysis
- Dobyns, W. B. (1996). "Absence makes the search grow longer". American Journal of Human Genetics 58 (1): 7–16. PMC 1914936. PMID 8554070
- Duncan-Johnson, C.C.; Donchin, E. (1977). "On quantifying surprise: The variation of event-related potentials with subjective probability". Psychophysiology 14 (5): 456–467.
- Evans, J. St. B. T.; Barston, J. L. & Pollard, P. (1983). "On the conflict between logic and belief in syllogistic reasoning". Memory and Cognition 11 (3): 285–306
- Evans,, J. St. B. T; Curtis-Holmes, J. (2005). "Rapid responding increases belief bias: Evidence for the dual-process theory of reasoning.". Thinking and Reasoning 11 (4): 382–389
- Fingelkurts AA, Fingelkurts AA, Rytsälä H, Suominen K, Isometsä E, Kähkönen S, Impaired functional connectivity at EEG alpha and theta frequency bands in major depression, BM-SCIENCE - Brain and Mind Technologies Research Centre, Espoo, Finland.
- Frith U, Frith CD. 2003, Development and neurophysiology of mentalizing, Philos Trans R Soc Lond B Biol Sci. 2003 Mar 29; 358(1431):459-73.

- Fukutake T et all., Late-onset hereditary ataxia with global thermoanalgesia and absence of fungiform papillae on the tongue in a Japanese family, Brain. 1996 Jun;119 (Pt 3):1011-21.
- Gao W, Lin W, Chen Y, Gerig G, Smith JK, Jewells V, Gilmore JH., 2009, Temporal and spatial development of axonal maturation and myelination of white matter in the developing brain
- Gilbert et al., 2007, Distinct regions of medial rostral prefrontal cortex supporting social and nonsocial functions, Soc Cogn Affect Neurosci. 2007 September; 2(3): 217–226.
- Gillberg C, Autistic children's hand preferences: results from an epidemiological study of infantile autism, 1983
- Graham Rawlinson, The Significance of Letter Position in Word Recognition PhD Thesis, 1976, Nottingham University
- Greengard, P, Mobley, P, Evidence for widespread effects of noradrenaline on axon terminals in the rat frontal cortex, Yale School of Medicine, 1984
- Gundersen GG, Cook T A, Microtubules and signal transduction, Curr Opin Cell Biol. 1999 Feb;11(1):81-94
- Gwen L. Schmidt, Casey J. DeBuse, Carol A. Seger, Right hemisphere metaphor processing? Characterizing the lateralization of semantic processes, , Department of Psychology, Colorado State University
- Hecht, S.; Shlar, S.; Pirenne, M.H. (1942). "Energy, Quanta, and Vision". Journal of General Physiology 25: 819–840. doi:10.1085/jgp.25.6.819. PMC 2142545. PMID 19873316
- Heneka et al., Locus ceruleus controls Alzheimer's disease pathology by modulating microglial functions through norepinephrine, Proc Natl Acad Sci U S A. 2010 March 30; 107(13): 6058–6063.
- Hoppenbrouwers, Arash Nazeri, Danilo R. de Jesus, Tania Stirpe, Daniel Felsky, Dennis J. L. G. Schutter, Zafiris J. Daskalakis, Aristotle N. Voineskos, White Matter Deficits in Psychopathic Offenders and Correlation with Factor

Structure, August 20, 2013DOI: 10.1371/journal.pone.0072375

- Indrani,G, SCIENCE; 313:1775-1781 (2006)
- Jan P. C. De Bruin et al., Orbital prefrontal cortex, dopamine, and social-agonistic behavior of male long evans rats
- Johnson, R.; Jr (1993). "On the neural generators of the P300 component of the event-related potential". Psychophysiology 30 (1): 90–97. PMID 8416066.
- Kempton et al., 2011, Structural Neuroimaging Studies in Major Depressive Disorder, Meta-analysis and Comparison With Bipolar Disorder – Arch Gen Psychiatry. 2011;68(7):675-690. doi:10.1001/archgenpsychiatry.2011.60
- Klauer, K.C.; J. Musch, B. Naumer (2000). "On belief bias in syllogistic reasoning". Psychological Review 107 (4): 852–884.
- Kleinhans NM et al., 2008, Abnormal functional connectivity in autism spectrum disorders during face processing.
- Kolb, B. and Whishaw, I. Q. (1990). *Fundamentals of Human Neuropsychology*, chapter 21. Memory, W. H. Freeman and Company, third edition.
- Lakatos (1978). The Methodology of Scientific Research Programmes: Philosophical Papers Volume 1. Cambridge: Cambridge University Press
- Liu, Hongwei et al. (2004), Isolation of Araguspongine M, a New Stereoisomer of an Araguspongine/Xestospongin alkaloid, and Dopamine from the Marine Sponge Neopetrosia exigua Collected in Palau
- Livingstone, Margaret; Ronald Harris-Warrick, Edward Kravitz (1980) "Serotonin and Octopamine Produce Opposite Postures in Lobsters". Science 208 (4439): 76–79. doi:10.1126/science.208.4439.76. PMID 17731572.
- Lutzenberger, W., Elbert, T., Rockstroth, B. (1987). A brief tutorial on the implications of volume conduction for the

interpretation of the EEG. Journal of Psychophysiology, 33. S56.

- Lyness S A, Zarow C, Chui H C. Neuron loss in key cholinergic and aminergic nuclei in Alzheimer disease: a meta-analysis. Neurobiol Aging 2003. 241–23.23
- Martinez et al. Dopamine Type 2/3 Receptor Availability in the Striatum and Social Status in Human Volunteers. Biological Psychiatry, 2010; 67 (3): 275 DOI: 10.1016/j.biopsych.2009.07.037
- Matsuyama, S S et al., Hypothesis: microtubules, a key to Alzheimer disease, 1989
- Mayer, AM (2006). "Polyphenol oxidases in plants and fungi: Going places? A review". Phytochemistry 67 (21): 2318–2331.
- Morón JA et al., 2002, Dopamine uptake through the norepinephrine transporter in brain regions with low levels of the dopamine transporter: evidence from knock-out mouse lines. The Journal of Neuroscience, January 15, 2002, 22 (2):389–395
- Nieuwenhuis S, Cohen J D. , Aston-Jones G, Decision Making, the P3, and the Locus Coeruleus–Norepinephrine System, 2005
- Petra J J Baarendse et al., Early Social Experience Is Critical for the Development of Cognitive Control and Dopamine Modulation of Prefrontal Cortex Function
- Pineda, J.A.; Foote, S.L.; Neville, H.J. (1989). "Effects of Locus Coeruleus Lesions on Auditory Event-Related Potentials in Monkey". J. Neurosci 9 (1): 81–93. PMID 2563282.
- Rabbitt, Patrick (1997) (ed). Methodology of frontal and executive function Hove UK: Psychology Press
- Schacter, Daniel L. (2011). Psychology Second Edition. 41 Madison Avenue, New York, NY 10010: Worth Publishers. pp. 136–137. ISBN 978-1-4292-3719-2
- Schwaerzel M et al. (2003). "Dopamine and octopamine differentiate between aversive and appetitive olfactory memories in Drosophila"

- Seeman, Philip, Shitij Kapur, Schizophrenia: More dopamine, more D2 receptors, Proc Natl Acad Sci U S A. 2000 Jul 5; 97(14): 7673–7675. PMCID: PMC33999
- Squires NK, Squires KC, Hillyard SA. (1975). Two varieties of long-latency positive waves evoked by unpredictable auditory stimuli in man. Electroencephalogr Clin Neurophysiol. 38(4):387-401. PMID 46819
- Swick, D., Pineda, J. a, Schacher, S., & Foote, S. L. (1994). Locus coeruleus neuronal activity in awake monkeys: relationship to auditory P300-like potentials and spontaneous EEG. Experimental brain research. Experimentelle Hirnforschung. Expérimentation cérébrale, 101(1), 86–92.
- Tang, F; Tao, L; Luo, X; Ding, L; Guo, M; Nie, L; Yao, S (2006). "Determination of octopamine, synephrine and tyramine in Citrus herbs by ionic liquid improved 'green' chromatography".
- Turner, Katherine C, Frost, Leonard, Linsenbardt David, John R McIlroy1 and Ralph-Axel Müller, Atypically diffuse functional connectivity between caudate nuclei and cerebral cortex in autism, 2006
- Werhahn et al., Decreased Dopamine D2/D3-Receptor Binding in Temporal Lobe Epilepsy: An [18F]Fallypride PET Study, Blackwell, 2006
- Weyrer, Simon et al., Serotonin in Porifera? Evidence from developing Tedania ignis, the Caribbean fire sponge (Demospongiae), Memoirs of the Queensland Museum 44: 659-665. Brisbane. ISSN 0079-8835.
- Van Alstyne et al. (2006). "Dopamine functions as an antiherbivore defense in the temperate green alga Ulvaria obscura
- Wolff, J et al., 2012, Differences in White Matter Fiber Tract Development Present From 6 to 24 Months in Infants With Autism; the IBIS Network
- Yao WD, Spealman RD, Zhang J, Dopaminergic signaling in dendritic spines, Biochem Pharmacol. 2008 Jun 1;75(11):2055-69.

- Yavich, Forsberg, Karayiorgou, Gogos, Männistö 2007, Site-Specific Role of Catechol-O-Methyltransferase in. Dopamine Overflow within Prefrontal Cortex and. Dorsal Striatum. Journal of Neuroscience (impact factor: 7.11). 10/2007; 27(38):10196-209. DOI:10.1523/JNEUROSCI.0665-07.2007
- Yu, A. J.; Dayan, P. (2005). "Uncertainty, neuromodulation, and attention". Neuron 46 (4): 681–92. doi:10.1016/j.neuron.2005.04.026. PMID

www.ingramcontent.com/pod-product-compliance
Ingram Content Group UK Ltd.
Pitfield, Milton Keynes, MK11 3LW, UK
UKHW041921190726
13854UKWH00003B/1362

9 788192 355979